Interdisciplinary Applied Mathematics

Volumes published are listed at the end of the book.

Springer

New York
Berlin
Heidelberg
Barcelona
Hong Kong
London
Milan
Paris
Singapore
Tokyo

Interdisciplinary Applied Mathematics

Volume 19

Problems in engineering, computational science, and the physical and biological sciences are using increasingly sophisticated mathematical techniques. Thus, the bridge between the mathematical sciences and other disciplines is heavily traveled. The correspondingly increased dialog between the disciplines has led to the establishment of the series: *Interdisciplinary Applied Mathematics.*

The purpose of this series is to meet the current and future needs for the interaction between various science and technology areas on the one hand and mathematics on the other. This is done, firstly, by encouraging the ways that mathematics may be applied in traditional areas, as well as point towards new and innovative areas of applications; and, secondly, by encouraging other scientific disciplines to engage in a dialog with mathematicians outlining their problems to both access new methods and suggest innovative developments within mathematics itself.

The series will consist of monographs and high-level texts from researchers working on the interplay between mathematics and other fields of science and technology.

Marek Kimmel David E. Axelrod

Branching Processes in Biology

With 54 Illustrations

Marek Kimmel
Department of Statistics
Rice University
Houston, TX 77251
P.O. Box 1892
USA
kimmel@stat.rice.edu

David E. Axelrod
Department of Genetics
Rutgers University
Nelson Biolabs
904 Allison Road
Piscataway, NJ 08854
USA
axelrod@biology.rutgers.edu

Mathematics Subject Classification (2000): 60J80, 60J85, 92DXX

Library of Congress Cataloging-in-Publication Data
Kimmel, Marek, 1953–
Branching processes in biology / Marek Kimmel, David E. Axelrod.
p. cm. — (Interdisciplinary applied mathematics; 19)
Includes bibliographical references (p.).

1. Biology—Mathematics models. 2. Branching processes. I. Title. II. Interdisciplinary applied mathematics; v. 19.
QH323.5 .K53 2001
570′.1′5118—dc21 2001042966

Printed on acid-free paper.

Printed in the United States of America.

9 8 7 6 5 4 3 2 1

ISBN 978-1-4419-2958-7 e-ISBN 978-0-387-21639-3

Springer-Verlag New York Berlin Heidelberg
A member of BertelsmannSpringer Science+Business Media GmbH

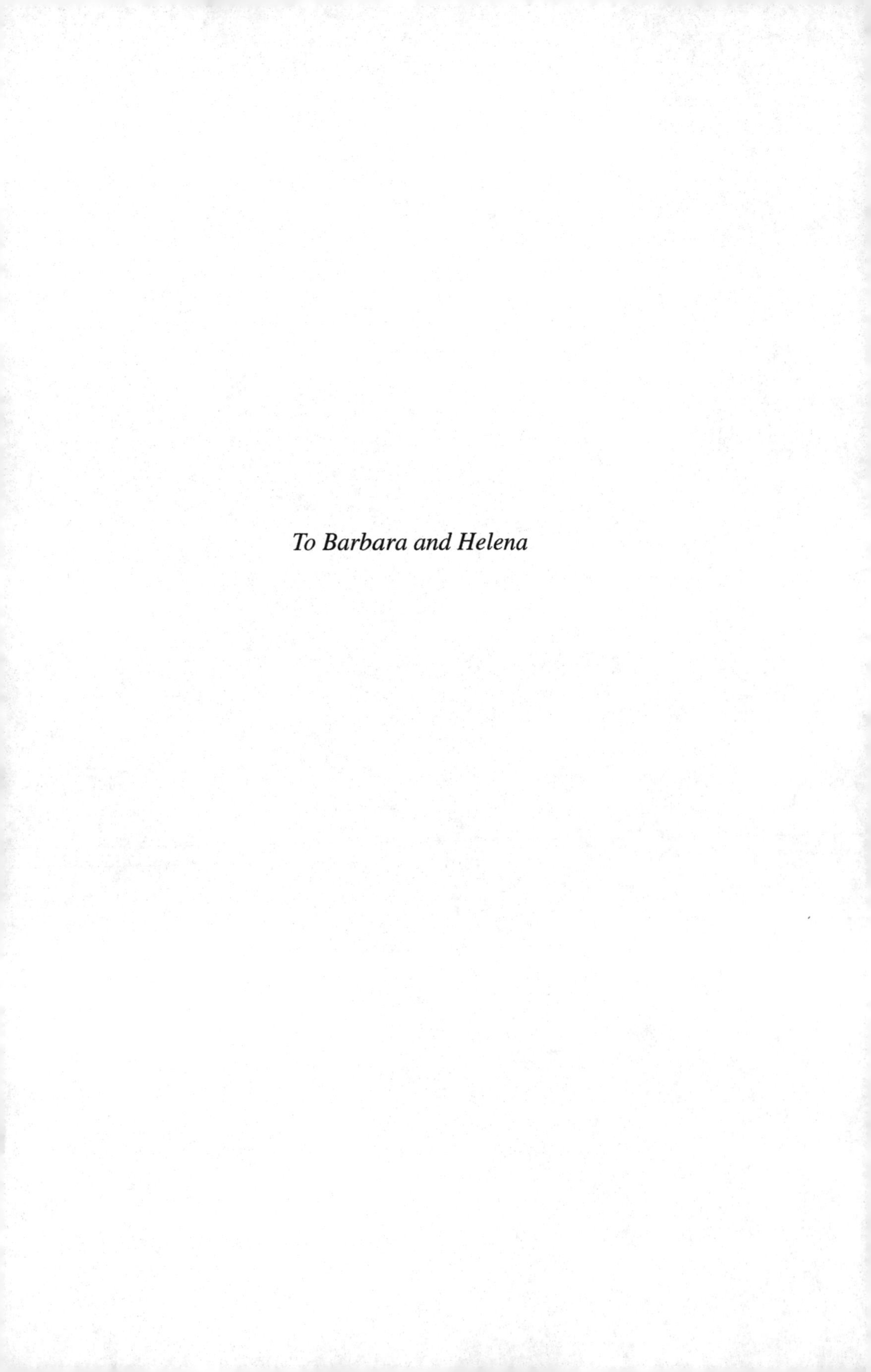

To Barbara and Helena

Preface

The theory of branching processes is an area of mathematics that describes situations in which an entity exists for a time and then may be replaced by one, two, or more entities of a similar or different type. It is a well-developed and active area of research with theoretical interests and practical applications.

The theory of branching processes has made important contributions to biology and medicine since Francis Galton considered the extinction of names among the British peerage in the nineteenth century. More recently, branching processes have been successfully used to illuminate problems in the areas of molecular biology, cell biology, developmental biology, immunology, evolution, ecology, medicine, and others. For the experimentalist and clinician, branching processes have helped in the understanding of observations that seem counterintuitive, have helped develop new experiments and clinical protocols, and have provided predictions which have been tested in real-life situations. For the mathematician, the challenge of understanding new biological and clinical observations has motivated the development of new mathematics in the field of branching processes.

The authors of this monograph are a mathematician and a cell biologist who have collaborated on investigations in the field of branching processes for more than a decade. In this monograph, we have collected examples of applications of branching processes from our own publications and from publications of many other investigators. Each example is discussed in the context of the relevant mathematics. We have made an effort to collect and review much of the published literature which has applied branching processes to problems in molecular and cellular biology, as well as selected examples from the fields of human evolution and medicine.

The intended audiences for this monograph are mathematicians and statisticians who have had an introduction to stochastic processes but have forgotten much of their college biology, and biologists who wish to collaborate with mathematicians and statisticians. Both audiences will find many examples of successful applications of branching processes to biological and medical problems. As an aid to understand the specific examples, we have provided two introductory chapters, one with background material in mathematics and the other with background material in biology, as well as two glossaries. As a didactic aid we have included problem sets at the end of Chapters 3, 4, and 5.

The book is organized as follows: Chapter 1 provides a mathematical background and motivating examples of branching processes. Chapter 2 provides an introduction to biological terms and concepts. The subsequent chapters are divided into specific areas of branching processes. Each of these chapters develops the appropriate mathematics and discusses several applications from the published literature. Chapter 3 discusses the Galton–Watson process, the oldest, simplest and best known branching process. Chapter 4 discusses the age-dependent process – Markov case, the time-continuous branching process with exponential lifetime distributions. Chapter 5 discusses the Bellman–Harris process, an age-dependent process. Chapter 6 gives a more systematic treatment of multitype processes, in which progeny may be of many types. Chapter 7 discusses branching processes with infinitely many types, stressing interesting properties which are different from the finite multitype situation. Appendices provide information on probability generating functions, construction of the probability space for the Bellman–Harris process, as well as a brief introduction to the Jagers–Crump–Mode process (the general branching process).

We have made an effort to broadly review the published literature on branching processes applied to biology. However, we had to select specific examples and we wish to apologize to our colleagues whose work has not been cited. We welcome comments from colleagues and students who are interested the field of branching processes.

A search of any university library or an Internet bookstore will reveal a number of volumes devoted to branching processes. Among the most important, we cite the fundamental books by Harris (1963) and by Athreya and Ney (1972). Multitype branching processes were first covered in the book by Mode (1971). General branching processes, in a systematic way, were explored by Jagers (1975). Each of these classics, particularly Jagers (1975), includes some biological applications. An important book concerning estimation of branching processes is by Guttorp (1991). The work by Asmussen and Herring (1983) involves a very mathematical approach. In addition, there exist at least a dozen or two of collections of papers and more specialized volumes. For example, Yakovlev and Yanev (1989) deal with cell proliferation models, mainly using branching processes. Recently, Pakes (2000) prepared a report on biological applications of branching processes, which is wider in scope (it has much on spatial branching and ecology, for example), but less detailed, although an area of overlap with our book exists. We believe that the scope of the present volume is unique in that it illustrates a paradigm, in which theoretical results are stimulated by biological applications and biological processes are illuminated by mathematics.

We gratefully acknowledge support from the following sources: National Institutes of Health, National Science Foundation, Keck's Center for Computational Biology at Rice University, New Jersey Commission on Cancer Research, Cancer Institute of New Jersey, Peterson Fund, Hyde and Watson Foundation, Glazer Family Fund, Rice University, Silesian Technical University, the University of Pau, and Rutgers University. Marek Kimmel was working on the final draft of this book while on a sabbatical leave at the Human Genetics Center at the University of Texas in Houston.

We thank Dr. William Sofer and Dr. Navin Sinha and several anonymous reviewers for helpful suggestions on the manuscript. Dr. Adam Bobrowski proofread the book for its mathematical correctness. His critical remarks improved it significantly. Remaining inaccuracies are our fault. Generations of graduate students at the Statistics Department at Rice University provided welcome feedback. Notably, David Stivers, Shane Pankratz, and Chad Shaw did their doctoral research in the area of branching processes. Professor Ovide Arino of the University of Pau, France, has contributed, over the years, to the understanding of connections between stochastic and deterministic population models. Professor Jim Thompson encouraged teaching this material at Rice and provided much constructive criticism. Professor Peter Jagers of Chalmers University in Gotheborg, Sweden, hosted Marek Kimmel on several occasions and provided much needed feedback. Our colleague Dr. Peter Olofsson was helpful with the more esoteric aspects of the theory. Our families showed warmth and patience during the gestation of this book. Jasiu helped with page proofs, and Kasia and Daniel did not hide words of advice.

We dedicate this book to our students, our teachers, and our families.

Houston, Texas, USA — Marek Kimmel
Piscataway, New Jersey, USA — David E. Axelrod

Contents

Guide to Applications, or How to Read This Book

As mentioned in the Preface, the book is organized by different classes of branching process, except for Chapter 1, providing general motivation and some mathematical background, and Chapter 2, providing biological background. Two glossaries at the end of the book give definitions of basic biological and mathematical terms commonly used in the book.

The inner structure of the book is a network of interconnected biological applications which increase in detail when modeled by progressively more sophisticated branching processes. The following list gives an overview of these applications:

- Cell cycle models

  Simplest version with death and quiescence, Section 3.2
  Two cell populations, Section 6.3
  Unequal division and growth regulation, Section 7.7.1
  Two types with growth regulation, Section 7.7

- Chemotherapy

  Stathmokinetic experiment, Section 5.4
  Cell-cycle-specific chemotherapy, Section 6.4

- Evolution theory

  Complexity threshold in early life, Section 3.4
  Galton–Watson branching process and macroevolution, Section 3.8.2
  Age of mitochondrial Eve, Section 4.4
  Yule's model of speciation, Section 7.8

- Gene amplification

  Galton–Watson branching process model, Section 3.6
  Stable gene amplification, Section 7.1
  Branching random-walk model, Section 7.4

CHAPTER 1

Motivating Examples and Other Preliminaries

The branching process is a system of particles (individuals, cells, molecules, etc.) which live for a random time and, at some point during lifetime or at the moment of death, produce a random number of progeny. Processes allowing production of new individuals during a parent individual's lifetime are called the general or Jagers–Crump–Mode processes (Fig. 1.1, top). They are suitable for the description of populations of higher organisms, like vertebrates and plants. Processes that assume the production of progeny at the terminal point of the parent entity's lifetime are called the classical processes (Fig. 1.1, bottom). They are usually sufficient for modeling populations of biological cells, genes, or biomolecules. In some processes, like the time-continuous Markov process, the distinction is immaterial because one of the progeny of a particle may be considered an extension of the parent.

One of the important notions in the theory of branching processes is that of the type space. The type space is the set, which can be finite, denumerable, or a continuum, of all possible varieties of particles included in the process. Particles of a given type may produce particles of different types. Restrictions on type transitions, as well as on the type space, lead to differing properties of resulting processes. The richness of these classifications is already apparent on the level of denumerable type spaces.

1.1 Some Motivating Examples

One of the oldest branching processes ever considered was the process in which "particles" were male individuals bearing noble English family names. An ancestor in such a process initiated a pedigree which might inevitably become extinct if all of the male descendants died without heirs. Is extinction of a noble family name inevitable in the long run? How many generations will elapse before extinction

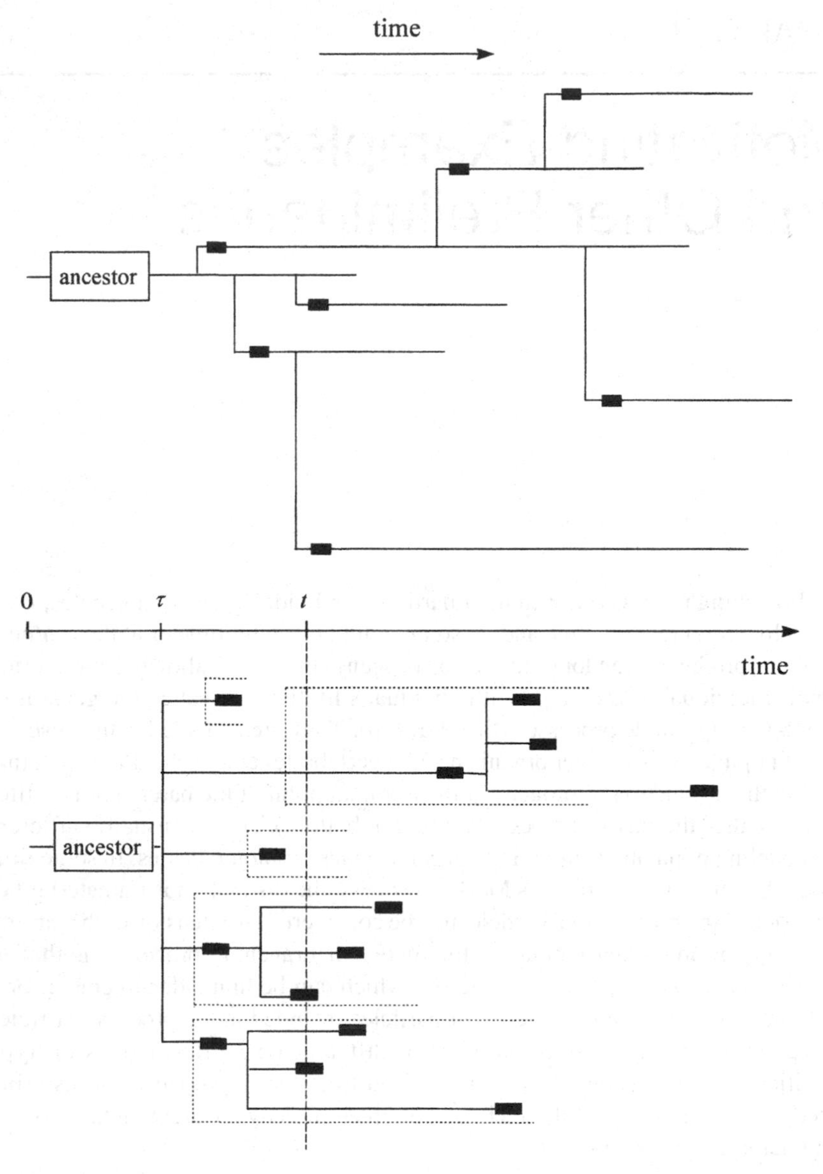

FIGURE 1.1. General (top) and classical (bottom) branching processes. Black rectangles depict individuals (objects, particles, etc.); horizontal lines depict lifetimes. Vertical lines are added to link individuals to their parents. The length of the vertical lines is arbitrary.

occurs? These are typical questions asked about a process in which the number of progeny of an individual may be equal to zero. Similar questions may be posed in a situation when a mutant cell initiates a small colony of precursor cancer cells. How likely are these colonies to die before they become numerous enough to overgrow the surrounding normal type?

A different type of question may be posed for processes in which the growth is assured by a sufficiently high proliferation rate. Then, the interesting parameter is the long-term growth rate and the size and composition of the population at a given time. This is typical of laboratory populations of biological cells, cultured with abundant nutrients and sufficient space. The same is true of prosperous individuals settling a large territory with few obstacles to growth. An interesting example is provided in the book by Demos (1982), in which it is stated that the average number of progenies surviving to maturity among the British colonists in New England in the seventeenth century was nine.

The patterns of branching may be quite complicated. An interesting example was given in the book by Harris (1963). In the course of evolution, new species are created by successful new varieties of organism which become reproductively separated from their ancestral species. This is an ordinary branching process. However, from time to time, an event occurs which creates a species so novel that it has to be considered an ancestor of a higher taxonomic unit than a species, a family. Therefore, branching becomes hierarchical: small particles (species) proliferate inside of large particles (families), which proliferate themselves, each started by a founder species. At both levels, extinction may occur. A similar branching pattern describes AIDS viruses proliferating in human T-lymphocyte cells. Divided lymphocytes inherit a portion of viruses present in the parent cell. If the number of viruses in a cell exceed a threshold, the cell dies. In this example, the two levels of branching compete with each other. Still another pattern is found in cancer cells inside which multiple copies of a gene increase a cell's resistance to treatment. If there are not enough of these, the cell becomes sensitive and dies.

1.2 *Application*: Polymerase Chain Reaction and Branching Processes

This section considers an important example of a branching process describing one of the most important tools of molecular biology, polymerase chain reaction (PCR). Following an introduction, we present a mathematical and simulation model constructed by Weiss and von Haeseler (1997). Material of this section is based on the Weiss and von Haeseler (1997) article, if not stated otherwise. Finally, we describe an application of PCR in artificial evolution.

1.2.1 Introduction to the mechanics of PCR

The following introduction has been adapted from a thesis by Shaw (2000). The PCR is an experimental system for producing large amounts of genetic material from a small initial sample. The reaction performs repeated cycles of DNA replication in a test tube that contains free nucleotides, DNA replication enzymes, primers and template DNA molecules. The PCR amplification technique operates by harnessing the natural replication scheme of DNA molecules, even using a naturally derived DNA polymerase protein. The result of the PCR is a vast amplification of a particular DNA locus from a small initial number of molecules.

Another feature of the PCR process is the stochasticity of amplification. Amplification is random because not every existing molecule is successfully replicated in every reaction cycle. Experimental evidence suggests that even the most highly efficient reactions operate at an efficiency around 0.8; that is, each double-stranded molecule produces an average of 0.8 new molecules in a given reaction cycle. The randomness in PCR can be attributed to the multiple molecular events which must occur in order to copy DNA.

The purpose of the PCR process is to produce clones (subpopulations with common descent) from DNA molecules from the small initial sample. Under ideal conditions, the molecules are identical in each clone, in the sense that the sequence of nucleotides A, T, C, and G in each molecule is either identical or complementary (A, T, C, and G replaced by T, A, G, and C, respectively) to the ancestral molecule of the clone (molecules in the initial samples may not be identical).

However, random alterations of nucleotides in DNA sequences, known as mutations, also occur during PCR amplification. In many PCR applications, mutations which occur during PCR hinder analysis of the initial sample, such as in the forensic setting. In other settings, however, PCR mutations are desirable, as is the case in site-directed mutagenesis studies and artificial evolution experiments (Joyce 1992). In both situations, analysis of variants generated during PCR is required. Interest focuses on the study of genetic diversity in a sample of molecules from the final stage of a PCR experiment. The molecules sampled are potentially related as descendants of a common ancestor molecule. The common ancestor of a family of PCR products is an initial molecule present at the starting stage of the amplification. The sampled molecules more commonly represent k samples of size 1 from distinct ancestor particles. This situation arises because PCR is performed from a very large number of initial molecules, usually more than many thousands. In either case, the genealogical method may be used to analyze the diversity of a sample taking into account the replication history and relatedness of the sampled molecules.

In order to assess the genealogy of the molecules in a sample, one must model the PCR and the structure of DNA replication. As in natural systems, DNA replication in the PCR is semiconservative, so that only one strand of each double-stranded DNA molecule is newly manufactured in a single replication event. Replication is semiconservative because each new single strand is built from a complementary antiparallel template strand during replication. Mutations can occur during construction of the new strand, so that newly fabricated strands may not be fully complementary to their templates. If a mutation occurs at some intermediate cycle of the PCR, the mutation will be propagated by the amplification procedure into all descendants of the mutant molecule. The goal is to study the sequence diversity of DNA molecules resulting from mutations during amplification.

1.2.2 Mathematical model

A model of PCR must include a model of the replication process and the mutation process. We use the single-stranded model, which is a simplification, because DNA is double-stranded [see a discussion in Shaw (2000)]. In the following, we frequently use molecule as a synonym for single-stranded sequence containing the subsequence of interest. Any other chemical molecules that are, in reality, present in a PCR tube are not considered. The replication process of PCR is described in terms of branching processes. The reaction proceeds through discrete cycles involving thermal and chemical processes. In each cycle, each single-stranded template should produce a copy. So, ideally, PCR is a binary fission process with discrete time (a special case of a Galton–Watson process; see Chapter 3). We assume that a PCR starts with S_0 identical copies of single-stranded sequences. Let S_i be the number of sequences present after the ith cycle. In cycle i each of the S_{i-1} template molecules is amplified independently of the others with probability λ_i. The probability λ_i can also be viewed as the proportion of amplified molecules in cycle i; hence, it is called the efficiency in cycle i. More precisely, the efficiency does not simply depend on the cycle number, but on the number of amplifications in the previous cycles and on PCR conditions.

If we assume that the random variable S_i depends only on λ_i and S_{i-1}, then the sequence $S_0, S_1, \ldots, S_i, \ldots$ forms a nonhomogeneous binary fission. If $\lambda_i = \lambda$ independently of the cycle number, then the accumulation of PCR product is a Galton–Watson branching process.

Because replication in PCR is not error-free, we add a mutation process to the model: We assume that a new mutation occurs at a position that has not mutated in any other sequence. Furthermore, we model the process of nucleotide substitution as a Poisson process with parameter μ, where μ is the error rate (mutation rate) of PCR per target sequence and per replication. This so-called infinitely many-sites model (ISM) does not allow for parallel or back mutations. In the case of PCR, this assumption seems reasonable because only a small number of mutations are observed in practice.

1.2.3 Genealogical approach

Computer simulations of stochastic processes have become a powerful tool to analyze data in situations where analytical methods are not feasible. In the population genetics literature, a prominent example is the coalescence process that describes the ancestral relationship between a sample of individuals in a population as one goes back in time (Tavaré 1984). Rather than trying to analyze the relations of all individuals in a population, the coalescent describes the (unknown) genealogy of a sample in terms of a stochastic process. If one starts with a sample of size n and traces back the ancestral history of these n lineages, one can compute the probability that at a time t, two lineages in the genealogy coalesce, that is, the most recent common ancestor (mrca) of the corresponding individuals is found. The probability depends on the sample size and the population trajectory. After a coalescent event occurs, the number of lineages is reduced by 1. The coalescent process stops when the mrca of the whole sample is found. In the situation of a stationary population of constant size, simple formulas are available that describe branch lengths in a genealogy of a sample, total length of a genealogy, and so forth. If one drops the assumption of constant size, it is more difficult to find analytical solutions, whereas it is still possible to get instructive results using simulation techniques.

The following simulation method to analyze PCR data bears similarity with the coalescent approach (see Section 4.3 for a more mathematical treatment of a similar process). In PCR, the offspring of the initial molecules are related by a randomly growing tree. Instead of generating this tree independently for each initial template, we adopt the following approach: For each initial molecule, the number of molecules in the PCR product in each cycle (the size trajectory) is computed (step 1 in the algorithm). Thereby, we distinguish two types of molecule: those that are immediate copies from a molecule of a previous cycle (filled circles in Fig. 1.2) and those that existed in the previous cycle (open circles in Fig. 1.2). From all molecules at the end of the PCR, a random sample of n sequences is

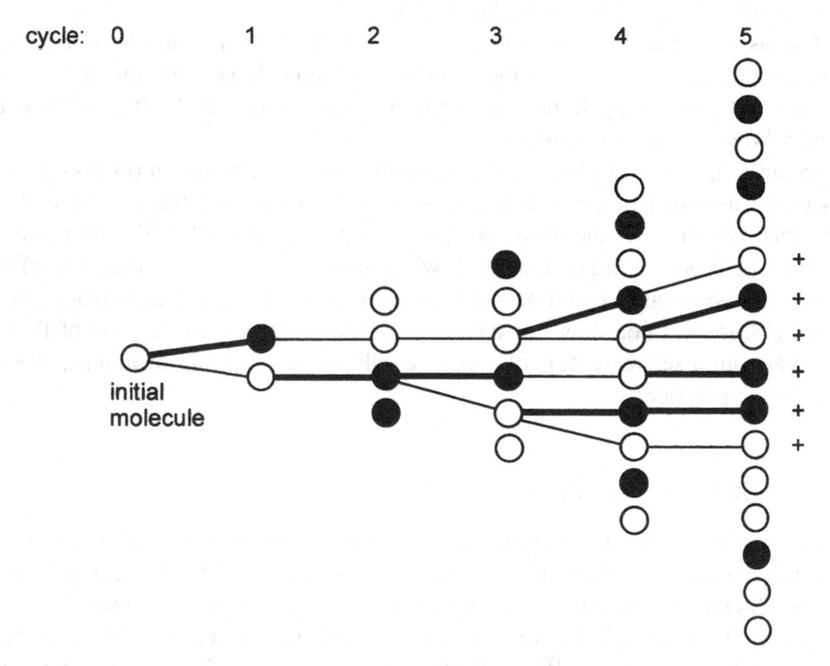

FIGURE 1.2. Graphical illustration of a subsample genealogy according to the example considered. Filled circles represent molecules that were newly amplified in a cycle; open circles represent molecules already present in the previous cycle. + indicates that the molecule is in the sample; thick lines represent a replication in the genealogy. Source: Weiss, G. and von Haeseler, A. 1997. A coalescent approach to the polymerase chain reaction. Nucleic Acids Research 25: 3082–3087. Figure 1, page 3084. Copyright: 1997 Oxford University Press.

drawn. Then, we randomly assign to each of the sampled sequences one of the initial molecules as ancestor (step 2) and regard the sets of sampled sequences that are descendents of the same initial molecule as subsamples. In the next step (step 3), we trace back the genealogies of all subsamples, separately.

Figure 1.2 illustrates this process for one initial molecule (and one subsample). In this example, we assume that a subsample of 6 sequences was drawn from a total of 16 sequences. In order to generate the subsample genealogy, the special features of PCR must be taken into account. A coalescent event (i.e., the merging of two molecules while going from cycle 5 to cycle 4) is only possible if exactly one of the two molecules is an immediate copy. Among the six sampled sequences, three were copied during cycle 5. Hence, at most three coalescent events are possible, and, in fact, one such event occurred. The coalescent process stops when only one molecule is left. If only one molecule is present and the cycle number is not equal to zero, then we have to determine how many replications took place from the initial molecule to this molecule.

After all subsample genealogies are generated, they are combined to one single genealogy (step 4). Finally, we superimpose a mutational process on the genealogy (step 5), where mutations are only allowed where replications took place in the genealogy (thick lines in Fig. 1.2). Before we describe the simulation more formally, we assign a number k, $k = 1, \ldots, S_0$, to each of the S_0 initial molecules.

1.2.4 Statistical estimation of the mutation rate

Weiss and von Haeseler (1997) carried out the estimation of the mutation rate μ, based on a published data set. They used a convenient measure of the diversity of the sample, resulting from PCR errors (mutations), defined as the number M_n of variable positions in a sample of size n (i.e., the number of the entries of the DNA sequence at which two or more variants are observed in the sample). In genetic literature, M_n is also known as the number of segregating sites.

Weiss and von Haeseler (1997) used the data of Saiki et al. (1988), who amplified a 239-bp region (i.e., a DNA sequence 239 nucleotides long) of genomic DNA. After $C = 30$ PCR cycles, $M_{28} = 17$ variable positions were observed when they sequenced $n = 28$ different clones. Furthermore, the authors measured the extent of amplification after 20, 25, and 30 cycles. They report an increase of 2.8×10^5, 4.5×10^6, and 8.9×10^6, respectively. This corresponds to an overall efficiency of 0.705 in 30 cycles. They also determined cycle-specific efficiencies from the data using the following formula:

$$\frac{\mathrm{E}(S_i)}{\mathrm{E}(S_{i-j})} = (1 + \lambda_i)^j, \quad i \geq j.$$

From the reported increase after 20, 25, and 30 cycles, they computed

$$\lambda_i = \begin{cases} 0.872, & i = 1, \ldots, 20 \\ 0.743, & i = 21, \ldots, 25 \\ 0.146, & i = 26, \ldots, 30. \end{cases}$$

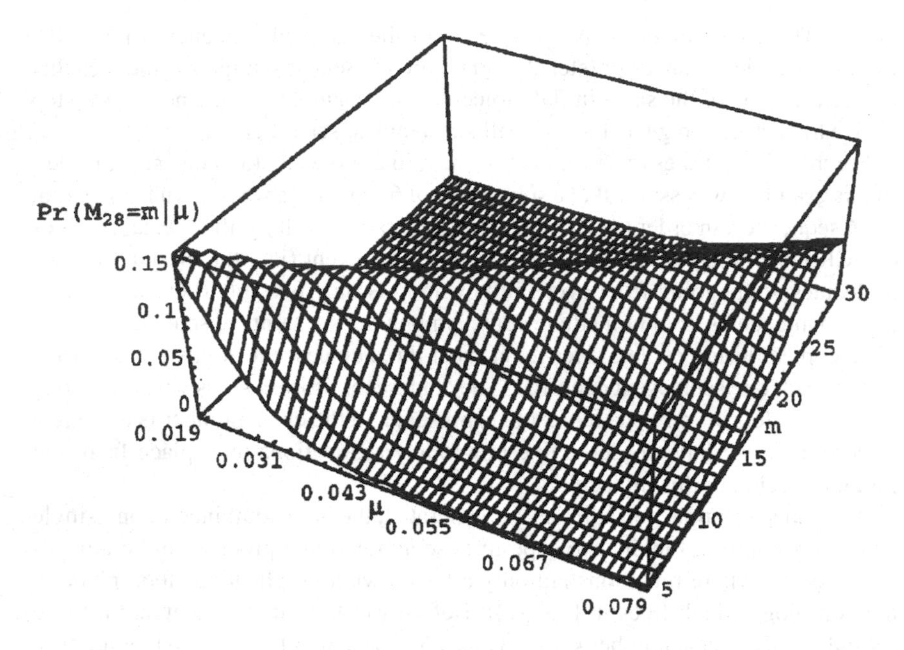

FIGURE 1.3. Example of simulated probability distributions $\Pr(M_n = m|\mu)$ for 100 equidistant values of μ ($S_0 = 100$). Source: Weiss, G. and von Haeseler, A. 1997. A coalescent approach to the polymerase chain reaction. Nucleic Acids Research 25: 3082–3087. Figure 2, page 3084. Copyright: 1997 Oxford University Press.

These values for λ_i were used in the simulations. Because no information about the number of initial molecules is given, the analysis was carried out for different S_0 values (1, 10, 100, 1000). They generated probability distributions $\Pr(M_n = m|\mu)$ for 100 equidistant values of μ in the interval [0.019, 0.079]. The scale on the m axis is limited to 30, because for the range of μ values considered, the likelihood has very small values for $m > 30$. If one takes the observed number of variable positions in the sample equal to 17 and cuts along this line through Figure 1.3, one gets $\text{lik}(\mu|M_{28} = 17)$, the likelihood function of μ given $M_{28} = 17$. Figure 1.4 shows the likelihood functions for $S_0 = 1$, 10, 100, and 1000. For each S_0, the position of the maximum of the likelihood function is the maximum likelihood estimate of μ.

1.2.5 *Mutagenic PCR and artificial evolution*

As mentioned earlier, mutations in the PCR may be desirable. One such example is provided by artificial evolution experiments, in which biomolecules, like RNA enzymes (ribozymes), are subjected to alternating rounds of amplification and mutation, and selection. In some classical experiments (Joyce 1992, Beaudry and Joyce 1992), high functional specificity of the evolved product was achieved. In these experiments, mutations provide the substrate for selection; therefore, un-

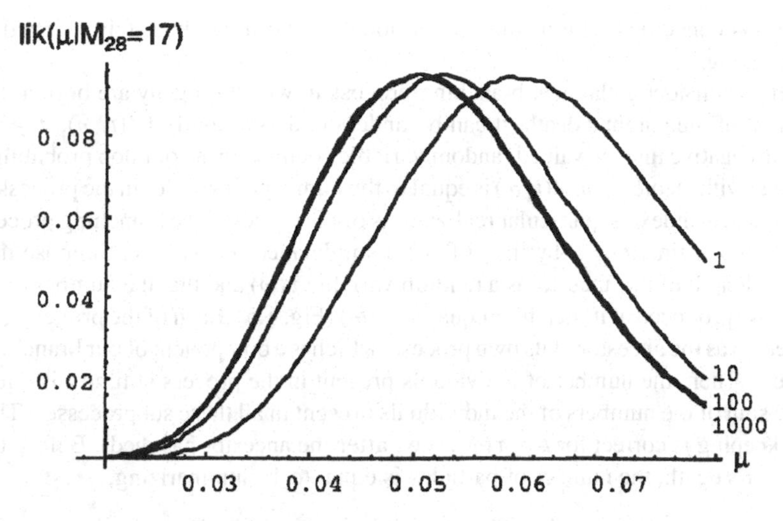

FIGURE 1.4. Simulated likelihood functions $\text{lik}(\mu|M_{28} = 17)$ for a published data set. The numbers in the graph represent the S_0 values used. Source: Weiss, G. and von Haeseler, A. 1997. A coalescent approach to the polymerase chain reation. Nucleic Acids Research 25: 3082–3087. Figure 3, page 3086. Copyright: 1997 Oxford University Press.

derstanding the mutational process in these experiments is very important. As of now, this remains an open problem.

1.3 The Branching Property

The branching property is a basic feature identifying processes studied in this book. It is responsible for many properties of the branching processes, some of them unexpected. The basic assumption involved is that each particle in the process behaves identically as all other particles and independently of all other particles (this latter is conditional on its existence). This may appear simple and obvious. However, consequences are far-reaching. Let us consider the clone, extending indefinitely into the future, originating from an ancestral particle. Such a clone is identical with the entire process we are studying. If we take any particle from this clone, then it gives rise to its own clone, which is a subprocess of the whole process. However, by the branching property, this subprocess is identical with the whole process. This realization provides a way to describe the process mathematically: It can be decomposed into subprocesses, which are identical (identically distributed, to be rigorous) with each other and with the entire process. In mathematical terms, branching processes belong to a class of stochastic objects called "self-recurrent" by Feller (1968, 1971).

Matters become a little more complicated if we allow particles of different types. The clones created by particles of different types are different, so the bookkeeping becomes more involved. However, the principle stays the same. The rest of this section is concerned with mathematical notation and it can be safely skipped at first reading.

Let us consider a classical branching process in which progeny are born at the moment of the parent's death. It can be understood as a family $\{Z(t, \omega),\ t \geq 0\}$ of non-negative integer-valued random variables defined on a common probability space Ω with elements ω. $Z(t, \omega)$ is equal to the number of particles in the process at time t, and ω index the particular realizations of the process. The branching process is initiated at time $t = 0$ by the birth of a single ancestor particle. Suppose that the life length of the ancestor is a random variable $\tau(\omega)$ and that the number of its progeny (produced at its death) is equal to $X(\omega)$ (Fig. 1.5). Each of the progeny can be treated as the ancestor of its own process, which is a component of our branching process. Then, the number of individuals present in the process at time t is equal to the sum of the numbers of the individuals present in all these subprocesses. This bookkeeping is correct for $t \geq \tau(\omega)$, (i.e., after the ancestor has died). Before the ancestor's death, the number of particles is equal to 1. Summarizing,

$$Z(t, \omega) = \begin{cases} \sum_{i=1}^{X(\omega)} Z^{(i)}(t, \tau(\omega), \omega), & t \geq \tau(\omega) \\ 1, & t < \tau(\omega), \end{cases} \tag{1.1}$$

where $Z(t, \tau(\omega), \omega)$ denotes the number of individuals at time t in the process started by a single ancestor born at time $\tau(\omega)$, and the additional superscript (i) denotes the ith independent identically distributed (iid) copy. The self-recurrence (or branching) property is embodied in the statement that the processes initiated by the progeny of the ancestor are independent and distributed identically as the ancestor:

$$Z^{(i)}(t, \tau(\omega), \omega) \stackrel{d}{=} Z^{(i)}(t - \tau(\omega), \omega). \tag{1.2}$$

Substitution of expression (1.2) into Eq. (1.1) leads to a recurrent relation:

$$Z(t, \omega) = \begin{cases} \sum_{i=1}^{X(\omega)} Z^{(i)}(t - \tau(\omega), \omega), & t \geq \tau(\omega) \\ 1, & t < \tau(\omega), \end{cases}$$

which we will use repeatedly.

In a rigorous way, construction of a branching process proceeds from the specification of distributions of life lengths and progeny numbers of individuals, to the construction of the probability space Ω, to deriving a specific form of relationships (1.1) and (1.2). Based on a classical construction by Harris (1963), the procedure has been extended to most general processes. In our applications, the existence and form of the probability space and self-recurrent relationships of the type (1.1)–(1.2) will be obvious. Therefore, usually, we will drop from the notation the argument ω, although, implicitly, it is always present.

The self-recurrence characterizing the branching process is one of the two conceivable ways the process can be defined. It is based on decomposing the process

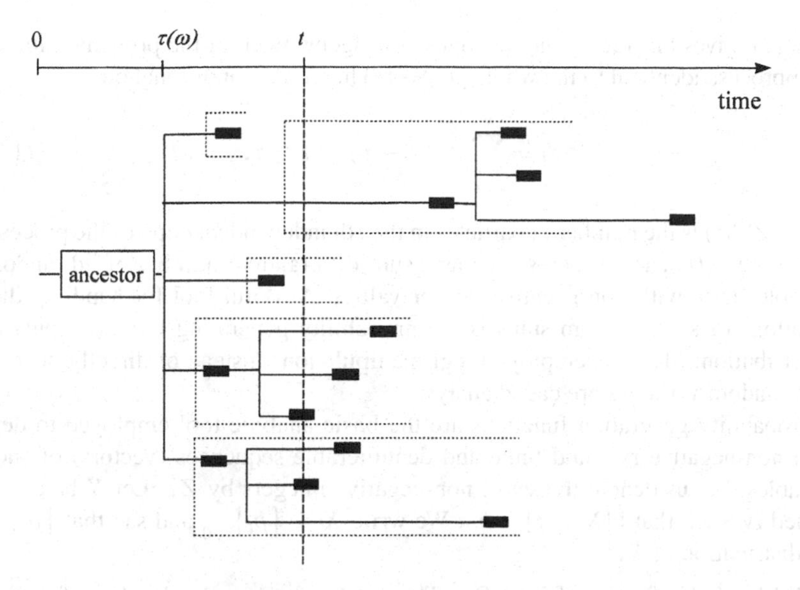

FIGURE 1.5. Decomposition of the branching process into subprocesses generated by the first-generation progeny of the ancestor; see Eq. (1.1). In the case depicted, the number of the first-generation progeny is equal to $X(\omega) = 5$. At time $t > \tau(\omega)$, the number of particles in the subprocesses generated by progeny 1, 2, 3, 4, and 5 is equal, respectively, to $Z^{(1)}(t, \tau(\omega), \omega) = 0$, $Z^{(2)}(t, \tau(\omega), \omega) = 1$, $Z^{(3)}(t, \tau(\omega), \omega) = 0$, $Z^{(4)}(t, \tau(\omega), \omega) = 3$, and $Z^{(5)}(t, \tau(\omega), \omega) = 3$. The total number of particles in the process at time t is seven.

into a union of subprocesses initiated by the direct descendents of the ancestor. It can be called the "backward" approach, in an analogy to the backward Chapman–Kolmogorov equations of Markov processes. A dual "forward" approach consists of freezing the process at time t, recording the states of all individuals at that time, and predicting their future paths (e.g., at $t+1$ or at $t+\delta t$). The backward–forward duality will be useful in some our considerations.

Branching processes have been widely used to describe growth and decay of biological populations. Their use has always overlapped with that of deterministic mathematical tools, like ordinary and partial differential equations. The doubtless applicability of branching processes is in studying small populations in which random fluctuations play a major role. However, some results concerning large populations are also easier to deduce using branching processes (see, e.g., Arino and Kimmel, 1993).

1.4 Probability Generating Functions and Analytical Methods

Consider a branching process composed of particles of one type. The number of particles at time t is denoted $Z(t)$. An ancestor is born at $t = 0$, and at random time τ, it gives birth to a random count of progeny. Each of the progeny initiates a subprocess identical to the whole process. Therefore, conditional on τ,

$$Z(t) \stackrel{d}{=} \sum_{i=1}^{X} Z^{(i)}(t-\tau), \quad t \geq \tau, \tag{1.3}$$

where $Z^{(i)}(t)$ is the number of particles in the ith independent copy of the process. Therefore, $Z(t)$ can be represented as a sum of a random number of iid random variables (rv), with non-negative integer values. A useful tool for handling distributions of such random sums is the probability generating function (pgf) of a distribution. Methods employing pgf manipulations instead of directly dealing with random variables are called analytic.

Probability generating functions are the basic analytic tool employed to deal with non-negative rv's and finite and denumerable sequences (vectors) of such variables. Let us denote the set of non-negative integers by Z_+. Let X be a Z_+-valued rv, such that $P[X = i] = p_i$. We write $X \sim \{p_i\}_{i \geq 0}$ and say that $\{p_i\}$ is the distribution of X.

Definition 1 (Definition of the pgf). The pgf f_X of a Z_+-valued rv X is a function $f_X(s) = E(s^X) = \sum_{i=0}^{\infty} p_i s^i$ of a symbolic argument $s \in U \equiv [0, 1]$. With some abuse of notation, we write $X \sim f_X(s)$.

The following theorem is a collection of results, which usually are given separately. All can be found, for example, in the book by Feller (1968). Also, see Pakes (1994).

Theorem 1 (The pgf theorem). *Suppose X is a Z_+-valued rv with pgf $f_X(s)$, which may not be proper. Let us denote by (N) the nontriviality condition $p_0 + p_1 < 1$.*

1. *f_X is non-negative and continuous with all derivatives on* $[0, 1)$. *Under (N), f_X is increasing and convex.*
2. *If X is proper, $f_X(1) = 1$; otherwise, $f_X(1) = P[X < \infty]$.*
3. *$d^k f_X(0)/ds^k = k!p_k$.*
4. *If X is proper, the kth factorial moment of X, $\mu_k = E[X(X-1)(X-2)\cdots(X-k+1)]$, is finite if and only if $f_X^{(k)}(1-) = \lim_{s\uparrow 1} f_X^{(k)}(s)$ is finite. In such case, $\mu_k = f_X^{(k)}(1-)$.*
5. *If X and Y are two independent Z_+-valued rv's, then $f_{X+Y}(s) = f_X(s) f_Y(s)$.*
6. *If Y is a Z_+-valued rv and $\{X^{(i)}, i \geq 1\}$ is a sequence of iid Z_+-valued rv's independent of Y, then $V = \sum_{i=1}^{Y} X^{(i)}$ has the pgf $f_V(s) = f_Y[f_{X^{(1)}}(s)]$.*
7. *Suppose that $\{X_i, i \geq 1\}$ is a sequence of Z_+-valued rv's. $\lim_{i\to\infty} f_{X_i}(s) = f_X(s)$ exists for each $s \in [0, 1)$ if and only if the sequence $\{X_i, i \geq 1\}$ converges in distribution to a rv X (i.e., if limits $\lim_{i\to\infty} P[X_i = k]$ exist for all k and are equal to $P[X_i = k]$, respectively). Then, $f_X(s)$ is the pgf of the limit rv X.*

The definition of the pgf can be generalized to the multivariate and denumerable cases (Appendix A).

Returning to the example in the beginning of this section, we notice that based on the pgf theorem, part 6, Eq. (1.3) can be now replaced by an equivalent pgf identity:

$$f_t(s) = f[f_{t-\tau}(s)], \quad t \geq \tau,$$

conditional on τ (i.e., given that the ancestor dies at age τ), where $f_t(s)$ denotes the pgf of Z_t and $f(s)$ denotes the pgf of X. As an example, let us consider the Galton–Watson process, which will be studied in detail in Chapter 3. In this process, all particles, including the ancestor, live for a fixed time equal to 1, so that $\tau \equiv 1$. This means that $f_t(s) = f\,[f_{t-1}(s)]$ for all $t \geq 1$, and so $f_t(s) = \underbrace{f\{f[\cdots f}_{n}{}_{t-n}(s)\cdots]\}$.

If we limit ourselves to integer t and notice that $f_0(s) = s$ (i.e., at $t = 0$, only the ancestor is present), then we obtain

$$f_t(s) = \underbrace{f\{f[\cdots f}_{t}(s)\cdots]\}, \tag{1.4}$$

which is the pgf law of evolution of the Galton–Watson process.

1.5 Classifications of the Branching Processes

1.5.1 Lifetime

The distribution of particle lifetime τ has much impact on the behavior and analysis of the process. As shown earlier, if $\tau \equiv 1$ (the Galton–Watson process, Chapter 3), it is enough to consider integer times. The pgf of Z_t [Z_t is an accepted notation for $Z(t)$ when time is discrete] is simply the t-fold functional iterate of the pgf of the progeny number, X [Eq. (1.4)].

Another important special case is when τ is distributed exponentially. The lack of memory of the exponential distribution leads to a process with continuous time which can be considered an interpolation of the Galton–Watson process between integer time points (Chapter 4).

Finally, if τ is an arbitrary non-negative random variable, the resulting process is called an "age-dependent" or Bellman–Harris process. It is more complicated to analyze than any of the two previous processes (Chapter 5).

1.5.2 Type space

The following is the list of the frequent variants of type space:

$$S = \begin{cases} \{1\}, & \text{single type} \\ \{1, \ldots k\}, & \text{multitype} \\ \{1, 2 \ldots\}, & \text{denumerable type} \\ R_+, R, [0, 1], & \text{continuous type} \\ \text{abstract.} & \end{cases}$$

The Galton–Watson and Bellman–Harris processes considered are single type but have their multitype, denumerable type and continuous type counterparts (Chapters 6 and 7). Abstract type spaces are used to create "superindividuals" composed of a number of original individuals and, in this way, to handle dependence among particles (Taïb, 1987).

1.5.3 Criticality

A very important classification is based on the mean progeny count $m = \mathrm{E}(X)$ of a particle. We introduce it using the example of the Galton–Watson process, but it is valid for all branching processes. By the pgf theorem, $\mathrm{E}(X) = f'(1-)$ and $\mathrm{E}(Z_t) = f_t'(1-)$. Differentiating the formula $f_t(s) = \underbrace{f\{f[\cdots f}_{t}(s)\cdots]\}$ with respect to s and substituting $s = 1$, we obtain using the chain rule of differentiation:

$$\mathrm{E}[Z_t] = f_t'(1-) = [f'(1-)]^t = m^t.$$

Therefore, in the expected value sense, the process grows geometrically if $m > 1$, stays constant if $m = 1$, and decays geometrically if $m < 1$. These three cases are called supercritical, critical, and subcritical, respectively:

$$\begin{array}{lll} m > 1, & \text{supercritical} \Rightarrow & \mathrm{E}[Z_t] \uparrow \infty, \\ m = 1, & \text{critical} \Rightarrow & \mathrm{E}[Z_t] = 1, \\ m < 1, & \text{subcritical} \Rightarrow & \mathrm{E}[Z_t] \downarrow 0. \end{array} \tag{1.5}$$

The above relationships are intuitively expected. However, the corresponding laws of extinction are less intuitive. Let us consider the probability $q_t = f_t(0) = \mathrm{P}[Z_t = 0]$ that the process is extinct at time t. We have $q_{t+1} \geq q_t$, as $Z_t = 0$ implies $Z_{t+1} = 0$. Because $0 \leq q_t \leq 1$ also, the sequence $\{q_t\}$ tends to a limit q which is the probability of eventual extinction. Moreover, because $f_{t+1}(s) = f[f_t(s)]$, then, setting $s = 0$, we obtain $q_{t+1} = f(q_t)$, and letting $t \to \infty$, $q = f(q)$. Therefore, q is the coordinate at which $f(s)$ intersects the diagonal. Let us note that $f(s)$ is convex and $f(1) = 1$. If $m = f'(1-) > 1$, then there exists $0 \leq q < 1$ such that $f(q) = q$. If $m = f'(1-) \leq 1$, then q has to be equal to 1. Therefore, we obtain that

$$\begin{array}{lll} m > 1, & \text{supercritical} & \Rightarrow q < 1, \\ m = 1, & \text{critical} & \left.\right\} \Rightarrow q = 1. \\ m < 1, & \text{subcritical} & \end{array} \tag{1.6}$$

The supercritical and subcritical processes behave as expected from the expression for the means. The critical process is counterintuitive. Although the mean value stays constant and equal to 1, the process becomes extinct almost surely ($q = 1$). This latter is only possible if the tail of the distribution is heavy enough to counterbalance the atom at 0. This suggests that a critical process is undergoing large fluctuations before it becomes extinct [cf, the discussion following Eq. (3.7)].

Further on, in Chapters 3–5 we will see that the limit behavior in all three cases may be characterized in more detail.

1.6 Modeling with Branching Processes

In this section, we discuss branching processes as a modeling tool, in a general and philosophical way. Our discussion owes a lot to ideas presented in a review article by Jagers (1991). We complement it with our own insights concerning the interactions between biology and branching processes.

As stated by Jagers (1991),

> Mathematical population theory is not the same as demography: Its object of study is not human populations. Nor is its object actual biological populations of, say, animals, bacteria or cells, or the physical populations of splitting particles in a cascade or neutron transport. Rather, its purpose is to study the common theme of these and many other empirical phenomena, an idealized pattern of free population growth, of sets changing as their members generate new set members.

For a mathematician,

> the essence of such a theory is mathematical in the same sense as geometry, the study of idealized shape. It is relevant for actual populations in so far as their reproduction is close to the idealized free reproduction and to the extent that this reproduction property is important for the evolution of the system as a whole. Thus in vitro cell kinetics is close to the pattern, at least if the population has enough nutrition and space, whereas the well-regulated growth of a couple of fetus cells into, say, a hand is dominated by features other than population growth.
>
> The population growth pattern is an important one, often playing a great role in the evolution of phenomena, and it can be discerned in many circumstances, ranging not only from demography to particle physics but including even data structures for sorting and searching in computer science or fractal sets arising in various types of mathematics. Sometimes the conclusions you can draw from the general mathematical study are even stronger than those obtained through more specialized models.

In opposition to the above, the approach advocated in our book is to explain biological observations in detail, the way mathematics is used in theoretical mechanics or relativity, and to generate predictions accurate enough to be practical. This approach may be considered a type of engineering. One may argue that by doing so, the modelers enter the turf reserved for other professions: biostatistics, demography, computer simulation, and biotechnology. Still, only the mathematical principles can explain the balance of factors contributing to the behavior of a population as a whole.

Unfortunately, this is not always appreciated in biology. One of the reasons is that much of modern biology is molecular biology. This latter, through the introduction of new techniques for gathering data and probing biological processes at a fundamental level, continuously provides an unprecedented amount of new information. Much of this information is connected only at a simplistic level (Maddox 1992). In the extreme reductionist view, everything can be reduced to a molecular switch which turns on or cuts off expression of a gene. In reality, it is frequently a delicate dynamic balance that creates a given behavior, and there might be alternative ways of inducing a biological system to display a seemingly related set of properties. For example, a complex human genetic disease, like diabetes, can arise through many alternative molecular pathways. A recent book by a well-known evolutionist discusses this subject (Lewontin 2000).

However, there are reasons to think that this situation soon may change. Mathematical and computational methods make steady headway in molecular biology. One example is the progress achieved in analysis of DNA sequences using hidden Markov models (Durbin et al. 1998). Recent sequencing of the human genome (Venter et al. 2001, International Human Genome Sequencing Consortium 2001) will generate a flood of research concerning expression of genes and relations between expression of different genes as well as the impact of these latter on evolution, human disease, and so on. Resulting problems will be difficult to answer without mathematics and/or computing power.

The subject of this book is the use of branching processes to model biological phenomena of some complexity, at different, though predominantly cellular or subcellular, levels. To understand the power and the limitations of this methodology, again we follow Jagers (1991). Probabilistic population dynamics arises from the interplay of the population growth pattern with probability. Thus, the classical Galton–Watson branching process defines the pattern of population growth using sums of iid random variables; the population evolves from generation to generation by the individuals getting iid numbers of children. This mode of proliferation is frequently referred to as "free growth" or "free reproduction."

The formalism of the Galton–Watson process provides insight into one of the fundamental problems of actual populations, the extinction problem, and its complement, the question of size stabilization: If a freely reproducing population does not die out, can it stabilize or must it grow beyond bounds? The answer is that there are no freely reproducing populations with stable sizes (see Section 3.3 for mathematical details). Population size stability, if it exists in the real world, is the result of forces other than individual reproduction, of the interplay between populations and their environment. This is true for structures much more general than the Galton–Watson process. For example, Breiman (1968, p. 98) demonstrates that the following is true: Consider a sequence of non-negative random variables $X_1, X_2, \ldots$, for which 0 is absorbing in the sense that $X_n = 0$ implies $X_{n+1} = 0$. Assume that there is always a risk of extinction in the following way. For any x, there is a $\delta > 0$ such that

$$\Pr[\text{there exists } n \text{ such that } X_n = 0 \mid X_1, \ldots, X_k] \geq \delta,$$

provided $X_k < x$. Then, with probability 1, either there is an n such that all $X_k = 0$ for $k \geq n$, or $X_k \to \infty$ as $k \to \infty$. So, the process either becomes extinct or grows indefinitely. We will consider more specific forms of this law for the Galton–Watson process (Theorems 2 and 3). Also, see Jagers (1992) for further discussion.

The next natural question is, what is the rate of the unlimited growth? It can be answered within the generation counting framework of Galton–Watson type processes (for biological examples, see Section 3.2). In a more general setup, we must know not only how many children parents get, but also the ages at child-bearings (even if they are constant and equal to 1, as in the Galton–Watson process case). In an even more general framework, the iid random variables describing reproduction have to be replaced by iid point processes, and the probabilistic addition of random variables by the superposition of point processes (Section C.1). In all these frameworks, in the supercritical case, when the average number of progeny of an individual is greater than 1, the growth pattern is asymptotically exponential. The parameter of this exponential growth is the famous Malthusian parameter. In the supercritical case, we can not only answer questions about the rate of growth but also questions about the asymptotic composition of nonextinct populations. What will the age distribution tend to be? What is the probability of being first-born? The average number of second cousins? What is the distribution of the time back to your nth grandmother's birth? Very important for biological applications, many of these questions do not have natural counterparts in deterministic models of unlimited growth.

Many other composition questions cannot be posed if we assume that all individuals are of one and the same type. Thus, we are naturally brought on to multitype branching populations: Whenever an individual is born, we know not only its parent's age but also its own type. This latter might be identified, in the most general case, with the individual's genotype.

Mathematically, the individual reproduction process then turns into a point process on the product space, type $\times$ age. Also, the evolution of the newborn's life will no longer be decided in an iid fashion but rather according to a probability kernel, determined by the type of the newborn. The introduction of various types of individual can be viewed as taking the step from independence to the simplest form of dependence in probability theory, the Markov dependence. One is born by one's parent, who decides when one is to come into this world and also passes on a genotype. Given these two inherited properties, one leads one's life independently of all of one's ancestors. This is the Markov model of population growth, the outcome of a straightforward combination of a vague population growth pattern with Markov dependence of random lives and reproduction.

Another general question is, what mathematical tools should be used to measure populations? Usually, we are interested in the number of individuals present at a given time. However, we might wish to count only individuals above a certain age. In some applications, we might be interested in the total number of individuals ever born. All of these variants are covered under the general concept of additive measures of population size, which goes back almost three decades (Jagers 1975). In this approach, each individual is measured by a random characteristic, a stochastic process, whose value at time t is determined by the individual's type, the individual's age now at time t, and the individual's, and possibly all of its progeny's, life careers. If the characteristic is assumed to vanish for negative ages, individuals are not taken into account before they are born. The measure of the population at time t is the sum of all of the characteristics, evaluated for all of the individuals as above. The simplest characteristic is the one that just records whether an individual is born or not, having the value of 1 if it is and 0 if it is not. It counts all individuals ever born.

As it will be seen, multitype branching processes are a tool for very detailed modeling using the Markov-multitype paradigm. In this setup, the type–space transitions become as important as branching itself. This is very clearly seen in processes with denumerable type spaces (e.g., the branching random walk describing gene amplification in Section 7.3). The process is a supercritical branching process, but the growth law is not Malthusian: Exponential growth is modified by a negative fractional power factor. Another example is the model of telomere shortening (Section 7.2). The transition law there is reducible, which produces a variety of unusual behaviors including polynomial growth.

A relatively new application of branching processes in genetics and evolutionism is the characterization of genealogies. In this approach, a sample of individuals from the process is considered at a given moment t, and conditionally on this information, distributions of past events related to the process are sought. The part of the process, existing before t, which contributed to the sample (i.e., excluding the individual whose descendants became extinct before t), is called the reduced process. Examples can be found in Section 4.3. The "backward-look" reduced-process limit laws for the critical and subcritical cases are quite different from those in the forward approach.

In classical population genetics models, the population size is a deterministic function of time. The stochastic part of the model is concerned with dependencies between the genetic makeups of the individuals. However, subpopulations of larger populations can be approximated by branching processes (Nagylaki 1990). This has important applications because various rare genotypes (e.g., mutant carriers of a rare genetic disease) belong to this category. One such application is in gene mapping (Kaplan et al. 1995) [i.e., determining the location of unknown genes based on their co-inheritance with known (marker) loci].

Branching processes are a conceptually simple tool for modeling diverse aspects of biological populations, not limited to demography, but reaching into cell and molecular biology, genetics, and evolution theory. They provide a framework for detailed considerations, allowing quantitative predictions, beyond metaphorical representations. With a future influx of detailed biological data, their importance for modeling is likely to increase.

CHAPTER 2

Biological Background

This chapter is a brief introduction, for mathematicians, to genes, cells and cancer. It provides general descriptions of the biological phenomena that are the subject of the mathematical applications developed in later chapters. More specific information relevant to each application is given at the beginning of the section containing the application. No knowledge of biology or chemistry is assumed beyond that learned in high school and forgotten due to disuse. Many biological details are omitted for lucidity. Readers familiar with the biological topics in this introduction may proceed directly to the later chapters.

2.1 Genomes: Changes in DNA and Chromosomes

2.1.1 Genome

The term "genome" refers to the molecules that function in the storage, expression, and inheritance of information in biological systems. The genome of humans and other organisms is dynamic. The number and sequence of its subunits can undergo rapid changes within a few generations.

2.1.2 DNA and genes

DNA (deoxyribonucleic acid) is the chemical that is the primary genetic material in the genome of all cells. It is responsible for the storage and inheritance of genetic information. DNA is a polymer consisting of two long complementary strands. Each strand contains a linear sequence of four monomer subunits called bases. The bases are abbreviated A, T, G and C. Each A base on one strand pairs with a T base on the other strand, and each G base pairs with a C base. The total length of DNA

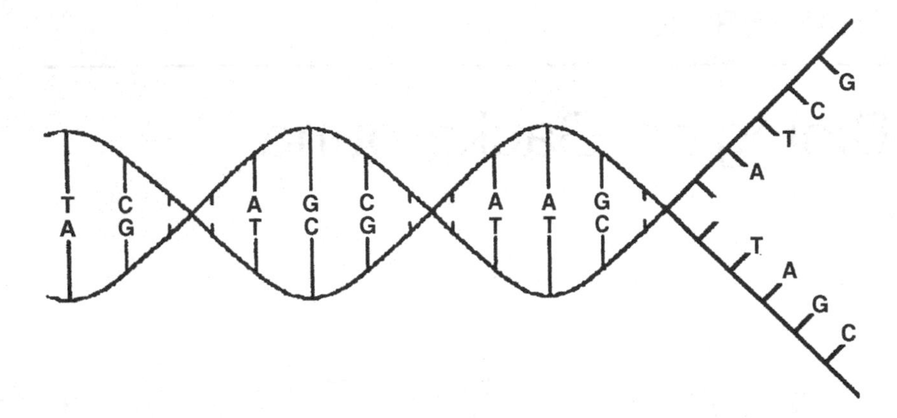

FIGURE 2.1. DNA structure. Pairs of complementary bases (A and T, G and C) hold together the double-stranded helix of DNA. The two strands are separated on the right, for replication, described further in Figure 2.3. The sequence of bases in DNA is transcribed into a sequence of bases in RNA and translated into a sequence of amino acids in protein which functions to determine the observable traits of the organism. Mutational changes in the sequence of bases in DNA result in changes in observable traits which are inherited.

in the genome of each mammalian cell is about 3×10^9 bases. Genes are specific subsequences of DNA that code for proteins. A given gene is typically 10^3–10^4 bases long and occurs one or only a few times in the genome. The mammalian genome contains approximately 30,000 different genes. The expression of the genetic information in DNA is accomplished by transcribing a sequence of bases in DNA into a sequence of bases into a related molecule, RNA (ribonucleic acid). The sequence of bases in RNA is then translated into a sequence of amino acids, the subunits of proteins. The proteins carry out catalytic and structural roles which result in the biological properties of cells and organs. Thus, the flow of information is usually as follows:

$$\text{DNA} \rightarrow \text{RNA} \rightarrow \text{protein} \rightarrow \text{phenotype};$$

in words,

$$\text{gene} \rightarrow \text{message} \rightarrow \text{catalyst} \rightarrow \text{observable trait}.$$

2.1.3 Mutation

An alteration in the sequence of bases in DNA is referred to as a mutation. The mutation may be as small as a change in a single base or as large as a rearrangement of most of the DNA in a chromosome. A mutation in DNA may result in an altered sequence of amino acids in protein and/or an altered amount of protein. This may result in a change in the ability of a protein to function properly, resulting in altered properties of cells and organisms. Many mutations are deleterious, but others may be advantageous or neutral. Examples of altered properties of cells containing mutations include the misregulation of cell growth and division leading to malignant tumors and the new capability of mutant cells to grow in the presence of a drug that would kill normal cells. A multitype process model describing mutations as occurring in several possible steps, and an improved method of estimating mutation rates from experimental observations are given in Section 6.1.

2.1.4 *Noncoding sequences of DNA*

Genes account for a small portion of the genome of mammals. Only about 5–10% of the DNA codes for proteins; the remainder is referred to as the noncoding portion of DNA. The function of the noncoding DNA is only partially understood. Some noncoding regions specify the sequence of bases in RNA that is never translated into protein. Another small portion consists of special DNA sequences that are required to maintain the ends of DNA molecules, called telomeres. Maintenance and loss of telomere sequences is discussed as a Bellman–Harris process with denumerable-type space in Section 7.2. Yet other noncoding sequences, centromeres, are required for the accurate segregation of the DNA into progeny cells. Fragments of DNA that do not contain centromeres distribute into progeny cells in uneven numbers. This is modeled as a single-type Galton–Watson branching process in Section 3.6 and as random walk with absorbing boundary in Section 7.4.

2.1.5 *Repeated sequences of DNA*

Much of the noncoding mammalian DNA consists of sequences which are repeated many times in the genome, most of unknown function. Some repeated sequences are tandemly distributed; others are dispersed throughout the genome. The length (number of bases) of a tandemly repeated unit may be as short as 1 base or longer than 10^3 bases. The number of repeated units may be as small as 2 or larger than 10^2. An increase in the number of tandemly repeated units is referred to as amplification, and a decrease as deamplification. The emergence of periodicities of tandemly repeated sequences in DNA by recombination slippage is simulated by a discrete stochastic dynamical system by Baggerly and Kimmel (1995). Repeats may also arise by other mechanisms, as discussed in Section 3.7 or in Bat et al. (1997).

2.1.6 *Gene amplification*

Regions of DNA may undergo an increase in number (amplification) or decrease in number (deamplification). The regions of DNA that undergo amplification and deamplification may contain genes or contain no genes. Amplification and deamplification of regions of DNA containing genes can result in increases or decreases in the amounts of proteins necessary for cell functions. Overproduction of rate-limiting proteins may confer new properties on cells. For example, if the protein is involved in cell proliferation, the cells with an increased amount of this protein may grow as malignant tumors. As another example, if the protein is the target of a toxic drug, then an increased amount of the protein may allow the cells to be resistant and grow in the presence of the drug. Models for gene amplification resulting in tumor cell growth and drug resistance are the subject of Sections 3.6, 7.1, and 7.4.

Some inherited human syndromes, such as predisposition to some cancers and neurological diseases, have been related to a rapid change in the numbers of copies of DNA sequences. An unusual aspect of these is the apparently explosive increase in the numbers of copies of some sequences from one generation to another. This increase has been modeled as an iterated Galton–Watson process in Section 3.7. In contrast to these cases of concerted increases in gene copy numbers, there are situations in which the number of amplified genes is unstable and decreases. Unstable decreases in numbers of amplified genes are modeled as a subcritical Galton–Watson process in Section 3.6 and as a branching random walk with absorbing barrier in Section 7.4.

2.1.7 *Chromosomes*

In human cells, the DNA of the genome in divided into pieces of various lengths, each containing large numbers of different genes. Each piece of DNA is folded compactly and associated with proteins and RNA to form a chromosome. In human cells, there are 23 pairs of chromosomes. Each chromosome contains one double-stranded piece of DNA from end to end. The ends of chromosomes are called telomeres. They have special sequences and structures that are necessary for the

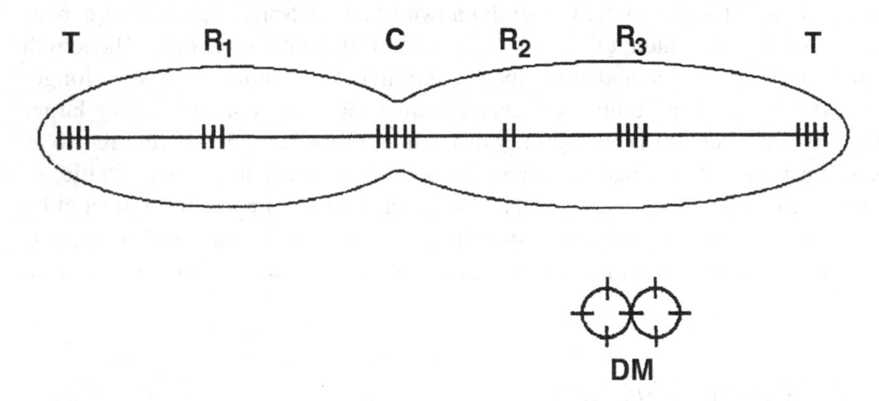

FIGURE 2.2. Chromosome. One double-stranded piece of DNA, represented here by a single horizontal line, extends from one end of a chromosome to the other. Several classes of repeated sequences of bases are represented. Telomere (T) repeats at each end of the chromosome function to maintain chromosome ends. Centromere (C) repeats function to separate chromosomes at mitosis and cell division. Other repeated (R) sequences of bases are dispersed throughout the chromosome. Some function to code for proteins (e.g., genes); others are noncoding sequences. Repeated sequences may exist as extrachromosomal elements, also called double minute (DM) chromosomes because of their appearance. They may replicate but are not partitioned evenly to progeny cells because they lack the centromeres of chromosomes. The number of repeated units may be variable.

replication of the end of DNA and the maintenance of chromosomes. DNA in chromosomes replicates and then the products separate from each other in a process called mitosis. Special DNA sequences near the center of chromosomes (centromeres) function to segregate one of each pair of duplicated chromosomes into each of two progeny cells during cell division. This process assures that each progeny cell receives one copy of each chromosome and its associated DNA-containing genes. Fragments of DNA without centromeres may increase in number (replicate) to more than two copies per cell, but without centromeres, there is no mechanism to distribute exactly equal numbers to each progeny cell.

2.1.8 DNA replication

DNA replication occurs by a so-called semiconservative mechanism. Two complementary parental strands of DNA separate and each strand forms a template for the production of a new complementary progeny strand. Usually, replication is initiated by the local separation of two strands, the replication fork, and then proceeds along the DNA. The result is two double-stranded DNA molecules, each molecule containing one old strand and one complementary new strand.

Two types of error in DNA replication have been proposed to result in amplification of repeated DNA sequences, replication slippage, and replication reinitiation. Replication slippage may occur when repeated DNA sequences on one strand fold

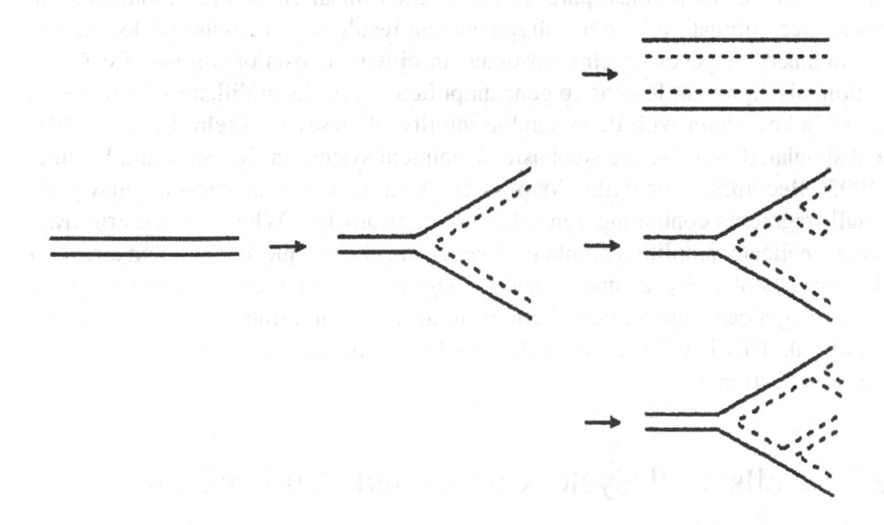

FIGURE 2.3. DNA replication. Double-stranded DNA (left) replicates by a semiconservative mechanism. The parental strands separate (center) and each codes for a new complementary strand. This results in two progeny double-stranded DNA molecules, each containing one old strand and one new strand (right top). Two types of error in DNA replication may result in locally repeated regions (repeat sequences). These errors include slippage and fold back, forming hairpin-like structures (right center), and replication reinitiation, forming branches within branches (right bottom).

back on themselves, forming a hairpin-like structure. This may cause slippage of the replication complex along one strand relative to the other, resulting in stuttering and repeated replication of a portion of the DNA sequence. The generation of unstable numbers of DNA repeats by replication slippage may contribute to the explosive increase in the numbers of repeated sequences in certain cancers and inherited neurological diseases. A mathematical model describing amplification of repeats by replication slippage has been developed (Bat et al. 1997), but it is not described as an application here because it is not a branching process. Replication reinitiation is another possible mechanism that may contribute to gene amplification. It is visualized as the start of a new replication fork before the previous replication fork has completed moving through the DNA. This leads to the formation of branched DNA structures. Gene amplification by replication reinitiation has been modeled in Section 3.7.

2.1.9 Recombination

Recombination is an exchange of pieces of DNA. Recombination can result in new combinations of genes and increases or decreases in the numbers of genes. Recombination occurs during the formation of germ cells for sexual reproduction (meiosis) and the division of nonsexual somatic body cells (mitosis). If replicated parts of chromosomes, called chromatids, align properly before recombination and exchange occurs, then new combinations of genes may occur and be segregated into sex cells. Sometimes, parts of chromatids misalign before recombination. Such a recombination with misalignment can result in an increase or decrease in the numbers of genes on chromosomes, in either meiosis or mitosis. Recombination misalignment leading to gene amplification or deamplification is modeled as a Markov chain with denumerable infinity of states in Axelrod et al. (1994) and simulated as a discrete stochastic dynamical system in Baggerly and Kimmel (1995). Recombination within loops of DNA on the same chromosome may yield small fragments containing genes but not centromeres. When the acentric fragments replicate, amplified numbers of genes may be produced in tandem arrays. If these pieces of DNA recombine and reintegrate into a larger chromosomal piece containing a centromere, then the tandem arrays of amplified genes can become stabilized. This is modeled as a Galton–Watson process with denumerable-type space in Section 7.1.

2.2 Cells: Cell Cycle Kinetics and Cell Division

2.2.1 Cells as the basic units of life

The basic structural unit of biological function and reproduction is the cell. Mammalian cells are in the range of 20×10^{-6} m in size, although there are many cells of different functions and different shapes that are smaller or larger. The structure of the cell is a series of bag-like compartments with specialized functions. The

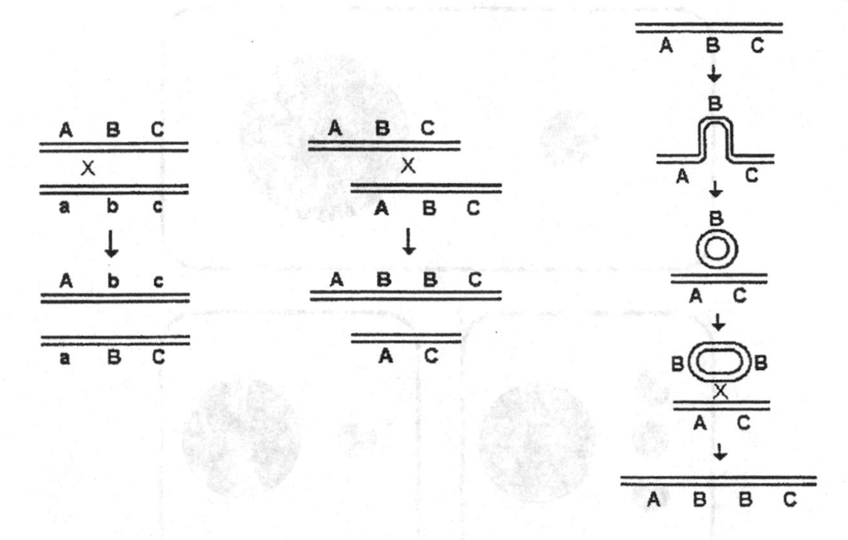

FIGURE 2.4. Recombination. New combinations and numbers of genes may be formed by rearrangement of pieces of DNA. Three examples are shown. Double-stranded DNA is represented by double lines, genes are represented by letters, and exchange is represented by an X. Left: The DNA molecules exchange genes, uppercase from the mother and lowercase from the father, to produce new combinations of genes which may then be passed on to the progeny. Center: The DNA strands slip and misalign before recombination, producing one molecule with an increased number of a gene and another molecule with a decreased number of the gene. Right: One molecule of DNA undergoes exchange with itself, producing a circular piece of DNA. If this piece replicates and then reintegrates, the result may be an increased number of a gene.

"bags" are made up of membranes that function as barriers and permit a selective transport of molecules. The innermost compartment is the nucleus which contains highly compacted DNA and accessory molecules for expression of genes. Outside of the nucleus is a series of compartments for the synthesis and degradation of molecules used for catalysis, structure, and energy generation. The outermost cell membrane and its accessory molecules also function as barriers and permit a selective transport, and, in addition, for communication with other cells. Communication between cells can occur via small molecules that diffuse between cells such as hormones or via molecules that become fixed to the surface of other cells, such as antigens which function in the immune system. A model for multivalent antigen binding as a multitype Galton–Watson process is given in Section 6.5.

2.2.2 *Cell growth and division*

Cells grow in size and divide into two. The DNA in the nucleus exactly doubles in amount, is packaged into chromosomes, and is then partitioned evenly between two progeny cells at cell division. However, other processes are less exact. The size to which cells grow before they divide is not exactly the same for all cells of

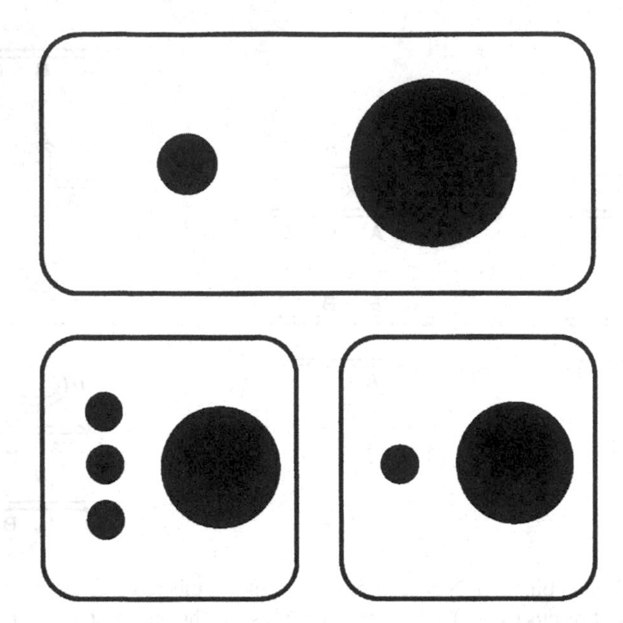

FIGURE 2.5. Cell division and partition of contents. During the growth of cells, the amount of DNA in the nucleus (large circle) doubles and is partitioned evenly into two daughter cells at cell division. However, other cell constituents may not exactly double and may not partition evenly, resulting in cells with different numbers of these constituents. These constituents include extrachromosomal pieces of DNA, subcellular organelles, and intracellular parasites.

a given type, the lifetimes of cells at division are not exactly the same for all cells, and the non-DNA materials are not partitioned exactly between the two progeny cells (see Figs. 2.5 and 2.6). The distribution of cell sizes and cell lifetimes may be stable over time for a population of one cell type, but differ for populations of cells of different types. Apparently, mechanisms exist to maintain these parameters within a population of cells of one type. Populations of cells with different values of parameters may differ in important characteristics, such as whether or not they are malignant. A Galton–Watson model describing the growth and division of cells is given in Section 3.2. Another model in the form of a Galton–Watson process with continuous-type space is described in Section 7.7.1.

During development of multicellular organisms, some cells divide into two cells which differ in shape and function. This situation is modeled as multitype branching process in Sections 6.3 and 7.7.2. If fragments of DNA are not connected to chromosomes, they may not exactly double in number before each cell division and may not partition exactly into the two progeny cells. Entities such as subcellular organelles or intracellular parasites can divide within dividing cells. An appropriate model for this is a Markov process model of infinitely many types. Such a model exhibits quasistationarity, as discussed in Section 7.5.

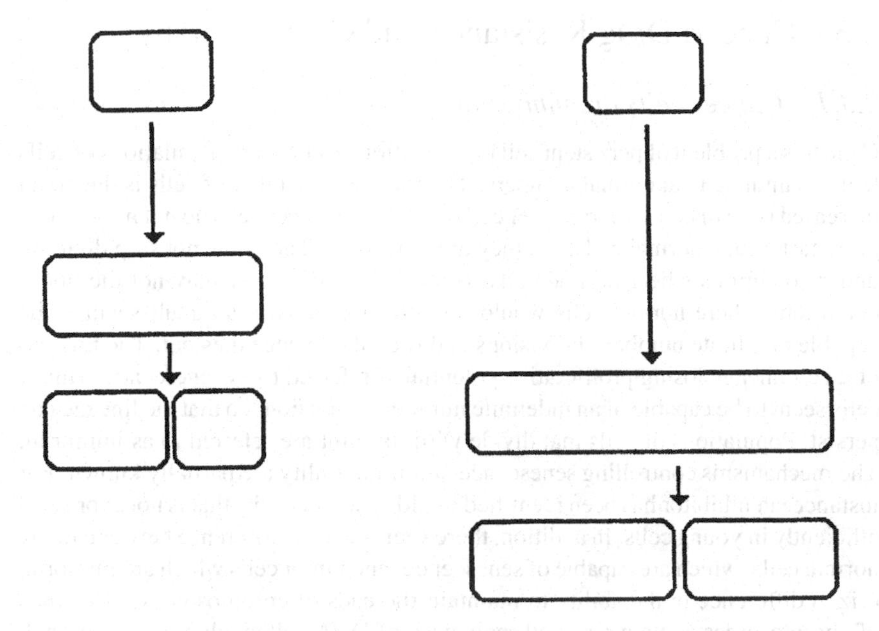

FIGURE 2.6. Cell division and cell size. Cells may grow for different times and attain different sizes before they divide. At cell division, cells may divide asymmetrically, resulting in progeny sibling cells of different size.

2.2.3 *Cell cycle kinetics*

The time period between cell birth and cell division is referred to as the cell cycle time. Several distinct events or phases can be distinguished during each cell's lifetime. The first event is the birth of two progeny cells at cell division, also called cytokinesis or mitosis, abbreviated M. The time gap between the birth of a new cell and the initiation of DNA synthesis is called gap one, abbreviated G_1. The period of DNA synthesis is abbreviated S. The time gap between S and the next mitosis is abbreviated G_2. After G_2, during the next M phase the cell divides to form two new cells. The sequence of phases M, G_1, S, G_2, and M repeats in progeny cells of each subsequent generation, thus the name cell cycle. For mammalian cells, a typical cell cycle time may be 12–24 h, or even longer. For a cell cycle time of 24 h, the duration for the cell cycle phases M, G_1, S, and G_2 might be 0.5, 8, 12, and 3.5 h. The duration of the G_1 phase is usually the most variable portion of the cell cycle. Cells which have longer cell cycle times, either because of genetics, environment, or developmental fate, usually have equally extended G_1 time periods, although important exceptions exist. The relative duration of the cell cycle phases in a growing population of cells can be inferred from the percentage of cells with different amounts of DNA or from the rate at which cells accumulate in one phase of the cell cycle when blocked with a phase-specific drug (stathmokinesis). Cell cycle kinetics are modeled as a Bellman–Harris process in Section 5.4 and as a Markov time-continuous branching process in Section 4.2.

2.3 Cancer: Drug Resistance and Chemotherapy

2.3.1 Cancer cells are immortal

Cancer is a problem of persistent cell proliferation. Tumors are populations of cells that accumulate in abnormal numbers. The increased number of cells is due to an increased ratio of cell birth rate over cell death rate. Cancer cells do not necessarily grow faster than normal cells, but they are persistent. They may not stop dividing under conditions where normal cells would stop and/or they may not die under conditions where normal cells would die. Normal cells in an adult seem to be capable of a finite number of divisions and then the lineage dies out. The process of a cell lineage losing proliferative potential is referred to as senescence. Tumor cells seem to be capable of an indefinite number of divisions so that the lineage can persist. Populations of cells that divide without limit are referred to as immortal. The mechanisms controlling senescence and immortality are partially known. For instance, an inhibitor has been identified in old senescent cells that is not expressed efficiently in young cells. In addition, there seems to be a difference between many normal cells, which are capable of senescence, and tumor cells which are immortal (viz. a difference in the ability to maintain the ends of chromosomes). The ends of chromosomes contain repeated sequences of DNA called telomeres. Although most of the length of DNA is duplicated exactly once during each cell cycle, that is not always true of the repeated DNA sequences in telomeres at the ends of chromosomes. The telomeric DNA sequences can increase or decrease in length at each round of DNA replication. Normal cells seem to progressively lose the telomeric repeat sequences and senesce (age), whereas some tumor cells seem to maintain them and continue to divide. This has been modeled as a Bellman–Harris process with denumerable-type space in Section 7.2.

2.3.2 Tumor heterogeneity and instability

Tumors are derived from single cells. This conclusion has resulted from observations in which all the cells in a tumor share a common change from normal cells. Cells from different tumors have different changes. The changes observed range in size from single-base mutations in DNA to large chromosome rearrangements. In addition to the common changes among the cells in a single tumor, many cells may show additional changes distinct for each cell in a tumor. In other words, tumors are monoclonal in origin but heterogeneous. Many tumors are genetically unstable, showing an increased probability of undergoing mutations or gene amplification. A mutant gene may produce a protein product with an altered function, and a gene with amplified number of copies may produce an increased amount of a protein. If the protein is the target of a toxic drug, then a tumor cell producing an increased amount of this protein may become resistant to this drug and escape effective chemotherapy. Gene amplification leading to drug resistance has been modeled using the Galton–Watson process in Section 3.6, modeled as a Galton–

Watson process with denumerable types in Section 7.1, and modeled as a branching random walk with absorbing barrier in Section 7.4.

2.3.3 *Cell cycle and resistance to chemotherapy*

Some forms of cancer chemotherapy attempt to exploit differences between the cell cycle of normal and malignant cells. For instance, a single drug that inhibits DNA synthesis would be expected to kill more tumor cells than normal cells, if more tumor cells than normal cells are synthesizing DNA during the period of chemotherapy. Sometimes, two or more drugs are used which affect different cell cycle phases, or have different mechanisms of inhibition (combination chemotherapy). The purpose is to overcome possible resistance to a single drug and to increase the probability of catching tumor cells in different phases of the cell cycle. Therefore, it is important to be able to determine the cell cycle phase durations of normal and malignant cells, the cell cycle phase specificity of drugs, and the changes that occur when cells are exposed to anticancer drugs. A multitype process is used to model changes in the cell cycle during chemotherapy, Section 6.4. The emergence of cross resistance (each cell resistant to two drugs) is modeled as a time continuous branching process in Section 4.2.

2.3.4 *Mutations in cancer cells*

Rates of mutations that occur in cancer cells are estimated by a method called the fluctuation test. The procedure was originally developed for bacteria. In this test, what is observed is the number of mutant cells arising in many parallel cultures, the number of cell divisions in the cultures, and the number of cultures which contain no mutant cells. The original method of calculating mutation rates from these observations is based on the assumption that each mutant cell resulted from a single rare event that is irreversible. This was appropriate for the bacterial mutations originally observed, but not for many mutations in cancer cells. The mutations may not be rare, irreversible, or due to a single step. Application of the fluctuation test to cancer cells required the development of methods that took into account these possibilities. Multitype branching processes were used to model two-step mutations and interpret data from the fluctuation test, see Section 6.1.

2.4 References

References cited here include textbooks and monographs on the topics discussed in this chapter. Specific citations to the primary literature are given in the applications sections of later chapters.

2.4.1 Textbooks and monographs in biology

Two outstanding textbooks in molecular and cellular biology used in upper level undergraduate and graduate classes are as follows:

- Alberts, B., Bray, D., Lewis, J. , Raff, M., Roberts, K., and Watson, J.D. 1994. Molecular Biology of the Cell, 3rd ed. Garland Publishing, New York.
- Lodish, H.F., Berk, A., Zipursky, S.L., Matsudaira, P., Baltimore, D., and Darnell, J.E. 1999. Molecular Cell Biology, 4th ed. Scientific American Books, W.H. Freeman, New York.

More detailed information on DNA and chromosomes can be found in the following:

- Kornberg, A. and Baker, T.A. 1992. DNA Replication, 2nd ed. W.H. Freeman, New York.
- Singer, M. and Berg, P. 1991. Genes & Genomes: A Changing Perspective. University Science Books, Mill Valley, CA.
- Wagner, R., Stallings, R.L., and Maguire, M.P. 1992. Chromosomes, A Synthesis. Wiley–Liss, New York.

Informative textbooks on cancer:

- Bishop, J.M. and Weinberg, R.A. 1996. Molecular Oncology. Scientific American Medicine, New York.
- Cooper, G.M. 1992. Elements of Human Cancer. Jones and Bartlett, Boston.
- Ruddon, R.W. 1995. Cancer Biology, 3rd ed. Oxford University Press, New York.
- Tannock, I.F. and Hill, R.P. (eds.). 1998. The Basic Science of Oncology, 3rd ed. McGraw-Hill, New York.

An excellent textbook on human molecular genetics:

- Strachan, T. and Read, A. 1999. Human Molecular Genetics, 2nd ed. Wiley, New York.

A few of the many Web sites with information on molecular and cellular biology include the following:

- Molecular Genetics Primer, Los Alamos,
  http://www.ornl.gov/hgmis/publicat/primer/intro.html
- Biology Hypertextbook, MIT
  http://esg-www.mit.edu:8001/esgbio/7001main.html
- WWW Virtual Library Biosciences, with many useful links,
  http://golgi.harvard.edu/biopages

2.4.2 Mathematical biology

Several textbooks provide examples of mathematical approaches to biological problems, although most examples are deterministic rather than stochastic models:

- Batschelet, E. 1992. Introduction to Mathematics for Life Sciences. 3rd edition. Springer-Verlag, New York.

- Brown, D. and Rothery, P. 1993. Models in Biology: Mathematics, Statistics and Computing. Wiley, New York.
- Hoppensteadt, F.C. and Peskin, C.S. 2001. Modeling and Simulation in Medicine and the Life Sciences. 2nd ed. Springer-Verlag, New York.
- Murray, J.D. 2002. Mathematical Biology: I An Introduction, 3rd ed. Springer-Verlag. New York.
- Murray, J.D. 2002. Mathematical Biology: II Spatial Models and Biomedical Applications. 3rd ed., Springer-Verlag, New York.
- Thompson, J.R. 1989. Empirical Model Building. Wiley, New York.

Monographs on mathematical biology include the following:

- Levin, S.A. (ed.). 1994. Frontiers in Mathematical Biology. Lecture Notes in Biomathematics, Volume 100. Springer-Verlag, Berlin.
- Segel, L.A. 1987. Modeling Dynamic Phenomena in Molecular and Cellular Biology. Cambridge University Press, Cambridge.
- Segel, L.A. (ed.). 1980. Mathematical Models in Molecular and Cellular Biology. Cambridge University Press, London.

For those who wish an introduction to stochastic process:

- Syski, R. 1988. Random Processes, A First Look, 2nd ed. Marcel Dekker, Inc., New York.
- Taylor, H.M. and Karlin, S. 1998. An Introduction to Stochastic Modeling, 3rd ed. Academic Press Harcourt Brace & Co., Boston.

A survey of some mathematical models of tumor cell growth, chemotherapy, and drug resistance that is useful for mathematicians and readable for biologists is

- Wheldon, T.E. 1988. Mathematical Models in Cancer Research. Adam Hilger, Bristol.

2.4.3 Arguments for mathematical modeling of biological phenomena

Several essays have emphasized the importance of mathematical modeling for progress in cellular and molecular biology:

- Goel, N.S. and Thompson, R.L. 1988. Models and their roles in biology, in Computer Simulations of Self-Organization in Biological Systems. Macmillan, New York, pp. 11–19.
- Huszagh, V.A. and Infante, J.P. 1989. The hypothetical way of progress. Nature 338: 109.
- Maddox, J. 1992. Is molecular biology yet a science? Nature 355: 201.
- Maddox, J. 1994. Cell-cycle regulation by numbers. Nature 369: 437.
- Maddox, J. 1995. Polite row about models in biology. Nature 373: 555.

CHAPTER 3

The Galton–Watson Process

The Galton–Watson process is the oldest, simplest and best known branching process. It can be described as follows.

A single ancestor particle lives for exactly one unit of time, and at the moment of death it produces a random number of progeny according to a prescribed probability distribution. Each of the first-generation progeny behaves, independently of each other, as the initial particle did. It lives for a unit of time and produces a random number of progeny. Each of the second-generation progeny behaves in the identical way, and so forth. From the fact that the life spans of all particles are identical and equal to 1, it follows that the process can be mathematically described using a discrete-time index, identical to the number of successive generation. The particle counts Z_n in the successive generations $n = 0, 1, 2, \ldots$ (where generation 0 is composed of the single initial particle) form a sequence of random variables with many interesting properties (the Markov property, for example). Properties of the Galton–Watson process provide intuitions about more complicated branching.

The simplicity of the Galton–Watson process makes it an appropriate and frequently employed tool for the introductory study of the processes of proliferation in biology. It is applicable whenever the hypothesis of discrete nonoverlapping generations is justified. An example of the Galton–Watson branching process is the process describing the polymerase chain reaction in Section 1.2.

3.1 Construction, Functional Equation, and Elementary Properties

The material in this section follows the style of Athreya and Ney (1972). Let us suppose that the number of progeny produced by each particle is a non-negative integer random variable with distribution function $\{p_k;\ k = 0, 1, 2, \ldots\}$.

3.1.1 *Backward equation*

Any particle existing in the process, except for the ancestor of the process, can be assigned to a subprocess traceable to a particular first-generation offspring of the ancestor. In other words, the process can be represented as a union of the subprocesses initiated by the first-generation offspring of the ancestor particle.

The number Z_{n+1} of particles in the generation $n+1$ of the process (or at time $n+1$) is equal to the sum of the particle counts in the generation n of all the Z_1 subprocesses initiated by the first-generation offspring of the ancestor particle. Let $Z^{(j)}_{1,n+1}$ denote the number of individuals at time $n+1$ in the process started by a single ancestor born at time 1. The additional superscript (j) denotes the jth iid copy. Mathematically, the random variable Z_{n+1} is equal to the sum of Z_1 random variables $Z^{(j)}_{1,n+1}$, or (see Fig. 3.1.)

$$Z_{n+1} = \begin{cases} Z^{(1)}_{1,n+1} + \cdots + Z^{(Z_1)}_{1,n+1}, & Z_1 > 0 \\ 0, & Z_1 = 0 \end{cases}$$

or

$$Z_{n+1} = \sum_{j=1}^{Z_1} Z^{(j)}_{1,n+1}. \tag{3.1}$$

Random variables $Z^{(j)}_{1,n+1}$ are *independent identically distributed* copies and their common distribution is identical to that of Z_n. Equation (3.1) can be equivalently written as

$$Z_{n+1} = \sum_{j=1}^{Z_1} Z^{(j)}_n.$$

By the pgf theorem 1 (part 6), it yields the following pgf identity:

$$f_{n+1}(s) = f_1[f_n(s)] = f[f_n(s)]. \tag{3.2}$$

If we note that $Z_0 = 1$ implies $f_0(s) = s$, this yields the following:

$$f_n(s) = f^{(n)}(s) = \underbrace{f\{\cdots f[f(s)]\cdots\}}_{n \text{ times}}; \tag{3.3}$$

that is, the pgf of Z_n is the nth functional iterate of the progeny pgf $f(s)$.

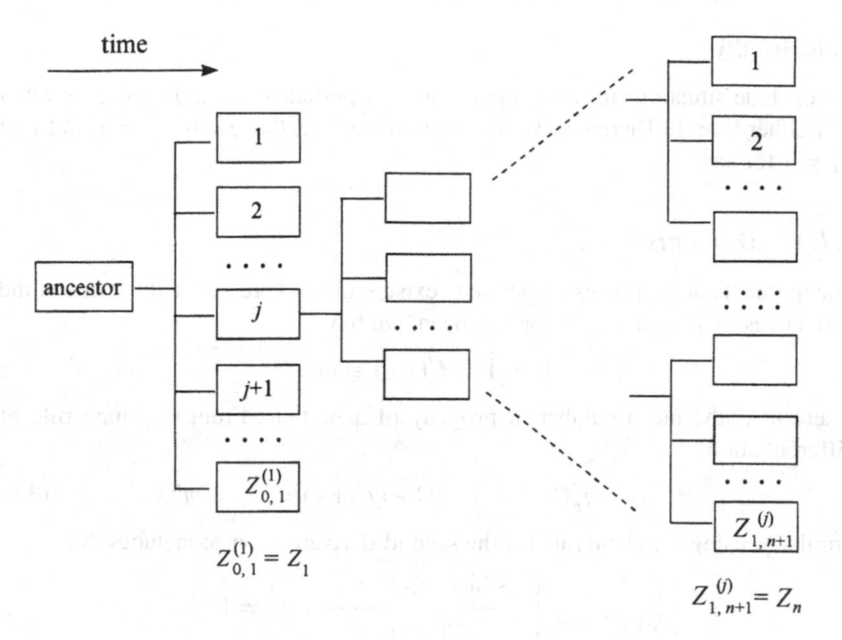

FIGURE 3.1. The backward equation for the Galton–Watson process.

3.1.2 *Forward equation*

An alternative approach is based on the fact that any particle in the $(n + 1)$st generation of the process can be traced to its parent in the nth generation of the process. Let X_{in} denote the number of progeny of the ith particle existing in generation n. More generally, let $\{X_{in}\}_{i\geq 1, n\geq 0}$ be a doubly infinite array of iid rv's such that $\mathrm{E}(X_{10}) = m < \infty$. Then,

$$\begin{aligned} Z_0 &= 1, \\ Z_{n+1} &= \begin{cases} X_{1n} + \cdots + X_{Z_n,n} & \text{if } Z_n > 0, \\ 0 & \text{if } Z_n = 0, \end{cases} \quad n \geq 1 \end{aligned} \tag{3.4}$$

or

$$Z_{n+1} = \sum_{i=1}^{Z_n} X_{in};$$

that is, the number of individuals (particles, cells), in the $(n + 1)$st generation of the process is equal to the number of progeny of all individuals in the generation n. In the terms of pgf's, we obtain a new recursion:

$$f_{n+1}(s) = f_n[f_1(s)] = f_n[f(s)]. \tag{3.5}$$

In the case of the Galton–Watson process, the above recurrence also leads to Eq. (3.3). However, for more general processes, the forward construction may not be feasible. We will return to this matter.

Nontriviality

We exclude situations in which the number of particles is deterministic or when it is either 0 or 1. Therefore, we assume throughout that $p_0 + p_1 < 1$ and that $p_j \neq 1$ for any j.

3.1.3 Moments

The moments of the process, when they exist, can be expressed in the terms of the derivatives of $f(s)$ at $s = 1$. For the mean, we have

$$\mathrm{E}(Z_1) = f'(1-) \equiv m,$$

where m is the mean number of progeny of a particle. From the chain rule of differentiation,

$$\mathrm{E}(Z_n) = f_n'(1-) = f_{n-1}'(1-)f'(1-) = \cdots = m^n. \tag{3.6}$$

Similarly, using the chain rule for the second derivative, one concludes that

$$\mathrm{Var}(Z_n) = \begin{cases} \dfrac{\sigma^2 m^{n-1}(m^n - 1)}{m-1}, & m \neq 1 \\ n\sigma^2, & m = 1, \end{cases} \tag{3.7}$$

where $\sigma^2 = \mathrm{Var}(Z_1)$ is the variance of the progeny count. Higher moments are derived similarly, if they exist. The linear growth of variance in the critical case ($m = 1$) is consistent with the "heavy tails" of the distribution of Z_n in the critical case as mentioned in Section 1.5.3.

3.1.4 The linear fractional case

Usually, after several iterations, the functional form of the iterates $f_n(s)$ becomes intractable. The linear fractional case is the only nontrivial example for which they have been explicitly computed. Suppose

$$p_0 = \frac{1-b-p}{1-p}, \quad p_k = bp^{k-1}, \quad k = 1, 2, \ldots, p \in (0, 1).$$

Then,

$$f(s) = 1 - \frac{b}{1-p} + \frac{bs}{1-ps} \tag{3.8}$$

and $m = b/(1-p)^2$. The equation $f(s) = s$ has roots q and 1. If $m > 1$, then $q < 1$; if $m = 1$, then $q = 1$; if $m < 1$, then $q > 1$. The following expressions are proved by induction (for a direct derivation, cf. Athreya and Ney 1972):

$$f_n(s) = 1 - m^n\left(\frac{1-q}{m^n - q}\right) + \frac{m^n[(1-q)/(m^n-q)]^2 s}{1 - [(m^n-1)/(m^n-q)]s}, \quad m \neq 1, \tag{3.9}$$

$$f_n(s) = \frac{np - (np + p - 1)s}{1 - p + np - nps}, \quad m = 1. \tag{3.10}$$

The linear functional pgf corresponds to the geometric distribution with a rescaled term at zero.

3.2 *Application*: Cell Cycle Model with Death and Quiescence

The material of this section follows Kimmel and Axelrod (1991). The fundamental step in the proliferation of a population of cells is the division of one cell into two cells. After completing its life cycle, each cell approximately doubles in size and then divides into two progeny cells of approximately equal sizes. Populations derived from single cells are referred to as clones or colonies. It has been experimentally observed that similar cells may not yield colonies with the same number of cells after the same amount of time. This may be due to various factors, like the randomness of cell death and quiescence.

3.2.1 The mathematical model

We consider a process more general than the standard Galton–Watson process. It is initiated by a single proliferating cell (Fig. 3.2). This cell divides, and each of its progeny, independently, may (i) become proliferative with probability (wp) p_2, (ii) become quiescent wp p_1, or (iii) die wp p_0. Quiescent cells continue to exist without proliferating or dying. They may return to active growth and

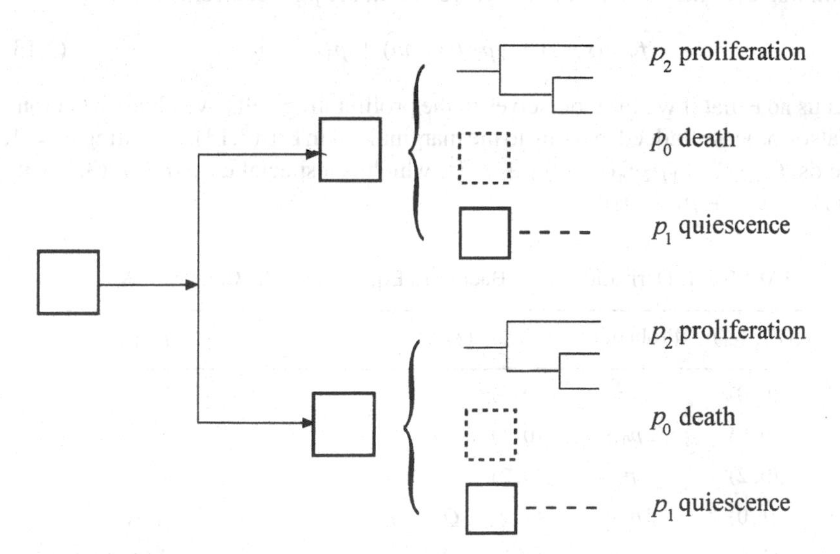

FIGURE 3.2. A schematic representation of the cell cycle model. Each of the daughter cells, independently, starts growing with probability p_2, dies with probability p_0, or becomes quiescent with probability p_1 ($p_0 + p_1 + p_2 = 1$).

proliferation even after a very long time, or eventually die; in the present model, these possibilities are not considered. We assume $p_2 + p_0 + p_1 = 1$.

The equations describing the model will be recurrences for the probability generating functions of the number of proliferating and quiescent cells present in the population in successive generation, analogous to the backward equation (3.1). Let us denote the number of proliferating cells in the nth generation by Z_n and the number of quiescent cells in the nth generation by Q_n. Also, let $Z_{n,k}^{(j)}$ and $Q_{n,k}^{(j)}$ denote respectively the number of kth generation proliferating and quiescent offspring of the jth of the Z_n proliferating particles of the nth generation. The number of offspring of a quiescent cell is always equal to 1, the same quiescent cell. We write the following equations:

$$Z_{n+1} = \sum_{j=1}^{Z_1} Z_{1,n+1}^{(j)}. \tag{3.11}$$

$$Q_{n+1} = \sum_{j=1}^{Z_1} Q_{1,n+1}^{(j)} + Q_1. \tag{3.12}$$

Let us denote by $f_n(s, w)$ the joint pgf of random variables Z_n and Q_n (see Definition 6 in the Appendix A). Let us note that $Z_{1,n+1}^{(j)}$ and $Q_{1,n+1}^{(j)}$ have distributions identical to those of Z_n and Q_n, respectively. To obtain the recurrence for the pgf, we first condition (Z_{n+1}, Q_{n+1}) on given values of (Z_1, Q_1). Table 3.1 lists all of the possibilities.

Multiplying the conditional values of $f_{n+1}(s, w)$ by their probabilities and summing over the rows of Table 3.1, we obtain the pgf recurrence:

$$f_{n+1}(s, w) = [p_2 f_n(s, w) + p_1 w + p_0]^2. \tag{3.13}$$

Let us note that if we limit ourselves to the proliferating cells, we obtain a Galton–Watson process. Indeed, passing to the marginal pgf in Eq. (3.13), by setting $w = 1$, yields $f_{n+1}(s) = [p_2 f_n(s) + p_1 + p_0]^2$, which is a special case of Eq. (3.2) with $f(s) = (p_2 s + p_1 + p_0)^2$.

TABLE 3.1. Derivation of the Backward Equation for the Cell Cycle Model

(Z_1, Q_1)	Probability	(Z_{n+1}, Q_{n+1})	$f_{n+1}(s, w)$
$(0, 0)$	p_0^2	$(0, 0)$	1
$(0, 1)$	$2p_0p_1$	$(0, 1)$	w
$(0, 2)$	p_1^2	$(0, 2)$	w^2
$(1, 0)$	$2p_2p_0$	$(Z_{1,n+1}^{(1)}, Q_{1,n+1}^{(1)})$	$f_n(s, w)$
$(1, 1)$	$2p_2p_1$	$(Z_{1,n+1}^{(1)}, Q_{1,n+1}^{(1)} + 1)$	$f_n(s, w)\, w$
$(2, 0)$	p_2^2	$(Z_{1,n+1}^{(1)} + Z_{1,n+1}^{(2)}, Q_{1,n+1}^{(1)} + Q_{1,n+1}^{(2)})$	$f_n(s, w)^2$

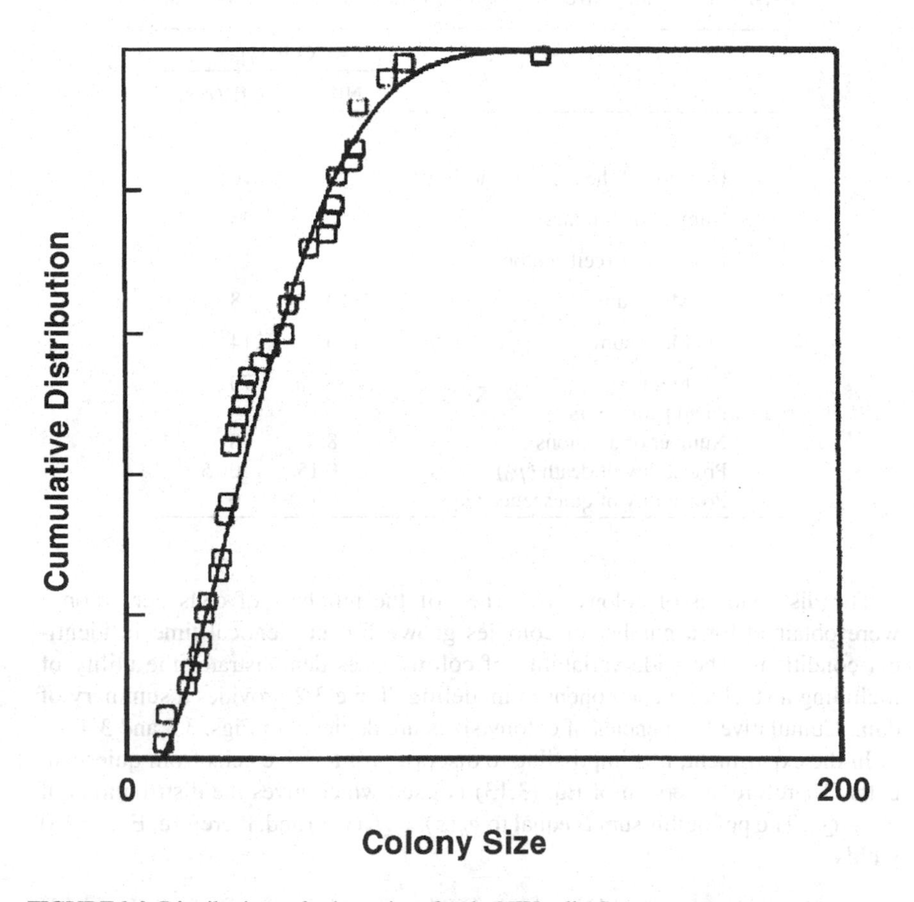

FIGURE 3.3. Distributions of colony sizes for the NIH cells. Squares represent experimental data and continuous lines have been generated by the model, as described in the text. The model satisfactorily reproduces the distributions of the NIH cells' colony sizes. Source: Kimmel, M. and D.E. Axelrod 1991. Unequal cell division, growth regulation and colony size of mammalian cells: A mathematical model and analysis of experimental data. Journal of Theoretical Biology 153: 157–180. Figure 3a, page 162. Copyright: 1991 Academic Press Limited.

3.2.2 Modeling biological data

In the article, Kimmel and Axelrod (1991), data on the colonies of cells have been modeled with the aid of Eq. (3.1). The data included empirical distributions of colony sizes of two varieties of cultured mouse fibroblast (connective tissue) cells. The first variety, the NIH cells, are relatively "normal" cells. The second variety was created by transferring the mutated *ras* oncogene, implicated in some malignant tumors, into NIH cells. The purpose of the experiments was to establish the differences in growth processes between the NIH and NIH(*ras*) cells.

TABLE 3.2. Colony Size Distribution: Data and Parameter Estimates

	Cell Type	
	NIH	NIH(*ras*)
Data		
Duration of the experiment (hours)	96	96
Number of colonies	52	45
Colony size (cells/colony)		
Minimum	10	8
Maximum	116	214
Median	33	70
Estimated parameters		
Number of divisions	8	8
Probability of death (p_0)	0.15	0.15
Probability of quiescence (p_1)	0.1	0

The distributions of colony sizes (i.e., of the numbers of cells per colony) were obtained for a number of colonies grown for an identical time in identical conditions. The wide variability of colony sizes demonstrates the utility of including a stochastic component in modeling. Table 3.2 provides a summary of data. Cumulative frequencies of colony sizes are depicted in Figs. 3.3 and 3.4.

In the experiment, it is impossible to discern proliferative cells from quiescent cells. Therefore, a version of Eq. (3.13) is used which gives the distributions of $Z_n + Q_n$. The pgf of this sum is equal to $g_n(s) = f_n(s, s)$ and, therefore, Eq. (3.13) yields

$$g_{n+1}(s) = [p_2 g_n(s) + p_1 s + p_0]^2. \tag{3.14}$$

This pgf is equal to $g_n(s) = \sum_j \pi_n(j) s^j$, where $\pi_n(j) =$P$\{Z_n + Q_n = j\}$ and Eq. (3.14) is equivalent to

$$\{\pi_{n+1}(j)\} = \{p_2\pi_n(j) + p_1\delta_{j1} + p_0\delta_{j0}\} * \{p_2\pi_n(j) + p_1\delta_{j1} + p_0\delta_{j0}\}, \tag{3.15}$$

where the asterisk denotes the convolution of distributions [i.e., $\{c(j)\} = \{a(j)\} * \{b(j)\}$ denotes $c(j) = \sum_{i=0}^{j} a(i)b(j-i)$, for $j = 0, 1, \ldots$]. The Kronecker symbol, δ_{jk}, is equal to 1 if $j = k$ and is equal to 0 if $j \neq k$. Recurrence (3.15), together with the condition $\pi_0(j) = \delta_{j1}$, $j \geq 0$, makes it possible to calculate the distributions of colony sizes.

During the time of the experiment, about $n = 8$ divisions occurred. Distributions $\{\pi_8(j),\ j \geq 0\}$ of colony size can be computed using different values of probabilities p_0 and p_1 of cell death and quiescence. The following strategy has been used:

1. Because the NIH(*ras*) cells have increased content of the mutated *ras* oncogene product, they are not likely to be quiescent; therefore, $p_1 = 0$ is assumed and

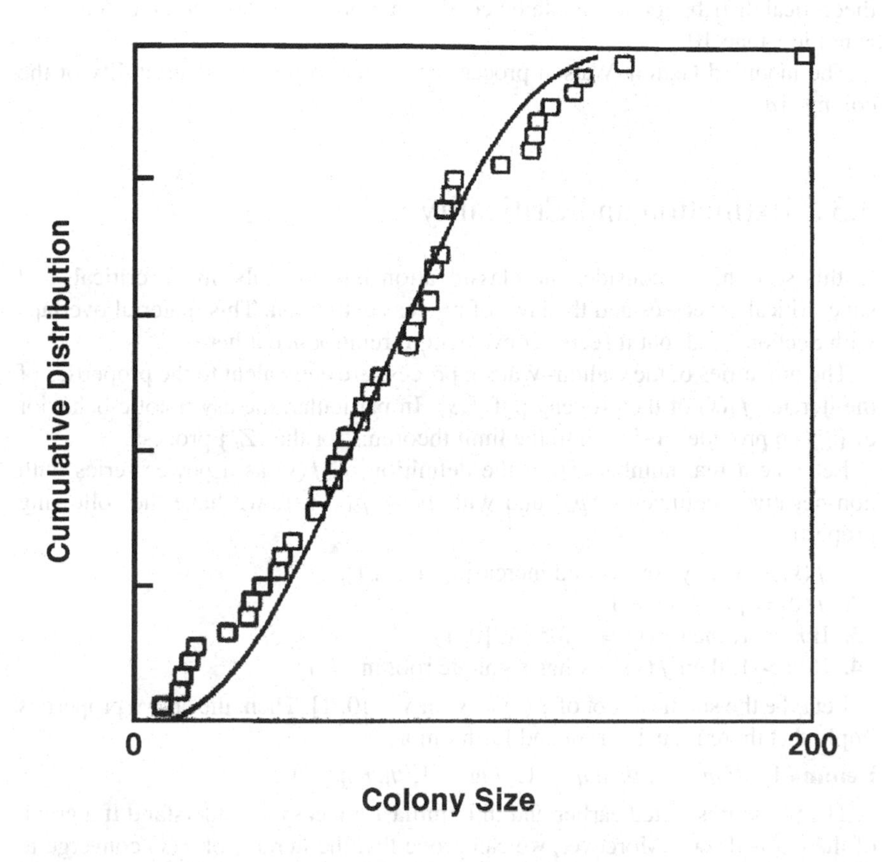

FIGURE 3.4. Distributions of colony sizes for the NIH (*ras*) cells. Squares represent experimental data and continuous lines have been generated by the model, as described in the text. The model satisfactorily reproduces the distributions of the NIH(*ras*) cells' colony sizes. Source: Kimmel, M. and Axelrod, D.E. 1991. Unequal cell division, growth regulation and colonysize of mammalian cells: A mathematical model and analysis of experimental data. Journal of Theoretical Biology 153: 157–180. Figure 3b, page 162. Copyright: 1991 Academic Press Limited.

the probability p_0 of cell death is varied to fit the empirical distribution, in the sense of least sum of squares of deviations of the model from the data. This gives $p_0 = 0.15$ (Fig. 3.4).

2. For the NIH cells, which are "normal," the same value ($p_0 = 0.15$) of the probability of cell death is used, but the probability p_1 of quiescence is varied until the distribution fits the data. This gives $p_1 = 0.1$ (Fig. 3.3).

As evident from Figure 3.3, the empirical cumulative frequency is initially steep, which suggests that colonies with less than 10 cells constitute a negligible fraction of the sample. These colonies were not counted in the experiment; therefore, the theoretical distribution is calculated conditional on the event that the colony size is not less than 10.

The modified Galton–Watson process accurately reproduces variability of the colony size.

3.3 Extinction and Criticality

In this section, we consider the classification into the subcritical, critical, and supercritical processes and the laws of process extinction. This material overlaps with Section 1.5.3, but it seems convenient to reintroduce it here.

The properties of the Galton–Watson process are equivalent to the properties of the iterates $f_n(s)$ of the progeny pgf $f(s)$. In particular, the asymptotic behavior of $\{f_n(s)\}$ provides insight into the limit theorems for the $\{Z_n\}$ process.

Let s be a real number. From the definition of $f(s)$ as a power series with non-negative coefficients $\{p_k\}$ and with $p_0 + p_1 < 1$, we have the following properties:

1. $f(s)$ is strictly convex and increasing in $[0, 1]$.
2. $f(0) = p_0$; $f(1) = 1$.
3. If $m \leq 1$, then $f(s) > s$ for $s \in [0, 1)$.
4. If $m > 1$, then $f(s) = s$ has a unique root in $[0, 1)$.

Let q be the smallest root of $f(s) = s$ for $s \in [0, 1]$. Then, the above properties imply that there is such a root and furthermore:

Lemma 1. *If $m \leq 1$, then $q = 1$; if $m > 1$, then $q < 1$.*

The properties stated earlier and in Lemma 1 are easy to understand if a graph of the pgf is drawn. Moreover, we can prove that the iterates of $f(s)$ converge to q.

Lemma 2. *If $s \in [0, q)$, then $f_n(s) \uparrow q$ as $n \to \infty$. If $s \in (q, 1)$, then $f_n(s) \downarrow q$ as $n \to \infty$. If $s = q$ or 1, then $f_n(s) = s$ for all n.*

As a special case of Lemma 2, we note that $f_n(0) \uparrow q$. However,

$$\begin{aligned} \lim_{n\to\infty} f_n(0) &= \lim_n \mathrm{P}\{Z_n = 0\} = \lim_n \mathrm{P}\{Z_i = 0, \text{ for some } 1 \leq i \leq n\} \\ &= \mathrm{P}\{Z_i = 0, \text{ for some } i \geq 1\} = \mathrm{P}\{\lim_{n\to\infty} Z_n = 0\}, \end{aligned}$$

which is, by definition, the probability that the process ever becomes extinct. Applying Lemma 1, we get the extinction probability theorem.

Theorem 2. *The extinction probability of the $\{Z_n\}$ process is the smallest non-negative root q of the equation $s = f(s)$. It is equal to 1 if $m \leq 1$, and it is less than 1 if $m > 1$.*

Theorem 2 states that the extinction probability depends on the parameter m, the mean progeny number of a particle.

Definition 2. If m is less than 1, equal to 1, or greater than 1, then the process is called subcritical, critical, or supercritical, respectively.

According to Theorem 2, the subcritical and critical processes eventually become extinct with probability 1. This is particularly surprising in the case of the critical process, for which the expected value of $\{Z_n\}$ stays constant. Therefore, some branching process models behave differently from their deterministic counterparts. Early in the history of the branching processes, it was remarked (Harris 1963), and then reiterated (Athreya and Ney 1972), that this instability of the Galton–Watson process is contrary to the behavior of biological populations, which tend to reach a state of balance with their environment. We will see, based on examples taken from cell and molecular biology, that the phenomena of extinction and instability do not contradict the rules of biology.

The Galton–Watson process is a Markov chain the state of which is equal to the number of particles present. We may classify its states into transient and recurrent. Recurrent states are revisited with probability 1. For transient states, this probability is less than 1. Let us examine the probability of not returning to a given state k. Let us denote by $P(k, j) = \mathrm{P}\{Z_{n+1} = j | Z_n = k\}$ the one-step transition probability of the process. The following is obtained:

$$\mathrm{P}\{Z_{n+i} \neq k, \text{ for all } i \geq 1 | Z_n = k\} \geq \left\{ \begin{array}{ll} P(k,0) = p_0^k, & p_0 > 0 \\ 1 - P(k,k) = 1 - p_1^k, & p_0 = 0 \end{array} \right\} > 0. \tag{3.16}$$

This is demonstrated as follows: If $p_0 > 0$, then one of the ways of not returning to state k is that the process becomes extinct in one step (this occurs with probability p_0^k). If this is impossible (i.e., if $p_0 = 0$), then we notice that one of the ways of not returning to k is to not return in a single step (if $p_0 = 0$, this occurs with probability $1 - p_1^k$). From Eq. (3.16), we deduce the following theorem.

Theorem 3. *All states except* $\{Z_n = 0\}$ *are transient; that is,*

$$\mathrm{P}\{Z_{n+i} \neq k, \text{ for all } i \geq 1 | Z_n = k\} > 0,$$

if $k \neq 0$. *In particular, this implies* $\lim_{n\to\infty} \mathrm{P}\{Z_n = k\} = 0$ *and* $\mathrm{P}\{\lim_{n\to\infty} Z_n = k\} = 0$, *for* $k \geq 1$. *The above, together with Theorem 2, implies that*

$$\mathrm{P}\{\lim_{n\to\infty} Z_n = 0\} = 1 - \mathrm{P}\{\lim_{n\to\infty} Z_n = \infty\} = q.$$

This latter property is known as the instability of the branching process.

3.4 *Application*: Complexity Threshold in the Evolution of Early Life

This example is taken from Demetrius et al. (1985). It concerns the ability of long biomolecules (polymeric chains) composed of smaller units (nucleotides) to replicate without error. The same problem can be considered in many different frameworks.

Let us consider a polymeric chain of ν nucleotides. If we assume that there is a fixed probability p that a single nucleotide is correctly copied, then the probability that a copy of the whole chain is correct is p^ν. Let us suppose that the chain replicates in a single time unit. During one generation step, the molecule either survives (with probability w) and produces a copy, which is accurate with probability p^ν, or it is destroyed with probability $1 - w$. A given molecule yields zero, one, or two molecules of the same type after one unit of time: The probabilities are $1 - w$, $w(1 - p^\nu)$, and wp^ν, respectively. The population of error-free molecules evolves according to a Galton–Watson branching process with pgf $f(s) = (1 - w) + w(1 - p^\nu)s + wp^\nu s^2$. This biomolecule is indefinitely preserved with a positive probability only if the process is supercritical [i.e., if $m = w(1 + p^\nu) > 1$], which yields

$$p^\nu > \frac{1 - w}{w}. \tag{3.17}$$

The probability of nonextinction is equal to $1 - (1 - w)/(wp^\nu)$.

Relation (3.17) implies that the error probability $1 - p$ yields a threshold for the length ν of the molecule, which does not become extinct with probability 1. If ν is larger than this threshold, then the molecular species becomes extinct.

3.5 Asymptotic properties

The limit theorems for the Galton–Watson process are important for many applications. Also, they suggest what to expect in more complicated processes. The limit laws are different in the supercritical, subcritical, and critical cases. In the supercritical case, the principal result is that the growth is asymptotically exponential, and that with probability 1, the random variable Z_n/m^n tends to a limit W. As a consequence, Z_n is approximated by Wm^n for large n. This is an extension of the exponential or Malthusian law of growth in the realm of stochasticity.

However, here the analogy ends. In the subcritical and critical cases, the probability of extinction is equal to 1 and the limit of Z_n/m^n is 0. Therefore, the "Malthusian law" is no longer a sensible approximation. In the subcritical case, it is replaced by the limit laws conditional on nonextinction (i.e., for the process $\{Z_n | Z_n > 0\}$). In the critical case, the limit distribution of $\{Z_n/n | Z_n > 0\}$ is exponential.

3.5.1 Supercritical process

The main mathematical fact used in this case is that the process $\{Z_n/m^n\}$ is a martingale.

Definition 3. A sequence of random variables $\{X_n, n \geq 0\}$ is called a martingale if $\mathrm{E}(|X_0|) < \infty$ and

$$\mathrm{E}(X_{n+1} | X_n, X_{n-1}, \ldots, X_1, X_0) = X_n.$$

Theorem 4. *If* $\{X_n, n \geq 0\}$ *is a non-negative martingale, such that* $\mathrm{E}(X_n) < \infty$ *for all n, then there exists a proper random variable X with finite expectation such that the following hold:*

1.

$$\lim_{n\to\infty} X_n = X, \quad \text{wp } 1.$$

2. *If the martingale is* L^2 *bounded [i.e., if* $\sup_n \mathrm{E}\,(X_n^2) < \infty$*], then the convergence occurs also in the* L^2 *sense. Then,* $\mathrm{Var}(X) = \lim_{n\to\infty} \mathrm{Var}(X_n)$.

Theorem 4 is a modified form of the theorem in Section 1.3 of the book by Neveu (1975).

If we set

$$W_n \equiv \frac{Z_n}{m^n},$$

then $\mathrm{E}(W_n) = 1$ and

$$\mathrm{E}(W_{n+1}|W_n) = m^{-(n+1)}\mathrm{E}(Z_{n+1}|Z_n) = m^{-(n+1)} m Z_n = W_n. \tag{3.18}$$

Because the Galton–Watson process is a Markov chain, then

$$\mathrm{E}(Z_{n+1}|Z_n, Z_{n-1}, \ldots, Z_1, Z_0) = \mathrm{E}(Z_{n+1}|Z_n).$$

An analogous property holds for the normalized process $\{W_n\}$; that is,

$$\mathrm{E}(W_{n+1}|W_n, W_{n-1}, \ldots, W_1, W_0) = \mathrm{E}(W_{n+1}|W_n).$$

Consequently, Eq. (3.18) demonstrates that $\{W_n\}$ is a martingale. Therefore, by part (1) of Theorem 4, we obtain the following.

Theorem 5. *If* $0 < m \equiv f'(1-) < \infty$*, then there exists a random variable W such that*

$$\lim_{n\to\infty} W_n = W, \quad \text{wp } 1.$$

In the critical and subcritical cases, $W \equiv 0$ since $q = 1$. Therefore, W might be nondegenerate only if $m > 1$. This is indeed true if an additional condition of finite variance of the number of progeny is imposed.

Theorem 6. *If* $m > 1$, $\sigma^2 < \infty$*, and* $Z_0 \equiv 1$*, then (i)* $\lim_{n\to\infty} \mathrm{E}(W_n - W)^2 = 0$*; (ii)* $\mathrm{E}(W) = 1$, $\mathrm{Var}(W) = \sigma^2/(m^2 - m)$*; (iii)* $\mathrm{P}\{W = 0\} = q = \mathrm{P}\{Z_n = 0$ for some $n\}$.

The Laplace transform $\phi(v) = \mathrm{E}(e^{-vW})$ of the distribution of W can be shown to satisfy a functional equation,

$$\phi(v) = f\left[\phi\left(\frac{v}{m}\right)\right], \tag{3.19}$$

the so-called Abel's equation. Indeed, relationship (3.2) can be rewritten in terms of Laplace transforms $\varphi_n(u) = \mathrm{E}[\exp(-uZ_n)] = f_n[\exp(-u)]$, where $u \geq 0$ is a symbolic argument,

$$\varphi_{n+1}(u) = f[\varphi_n(u)]. \tag{3.20}$$

Because the Laplace transform of the distribution of W_n is equal to

$$\phi_n(u) = \mathrm{E}[\exp(-uW_n)] = \mathrm{E}\left[\exp\left(-\frac{u}{m^n}Z_n\right)\right] = \varphi_n\left(\frac{u}{m^n}\right)$$

and, conversely, $\varphi_n(u) = \phi_n(um^n)$, substitution into Eq. (3.20) yields $\phi_{n+1}(um^{n+1}) = f[\phi_n(um^n)]$. After a change of variables $v = um^{n+1}$, we obtain

$$\phi_{n+1}(v) = f\left[\phi_n\left(\frac{v}{m}\right)\right].$$

Because $W_n \to W$ in distribution, $\phi_n(v) \to \phi(v)$ and the limit (which is the Laplace transform of the distribution of rv W) satisfies Abel's equation (3.19).

Example. In the linear fractional case of Section 3.1.4, the distribution of W can be directly calculated. Its "density" can be expressed as

$$f_W(w) = q\delta(w) + (1-q)^2 e^{-(1-q)w}, \quad \geq 0;$$

that is, it has an atom at 0 and the remaining part is negative exponential (cf. Section 3.10).

3.5.2 Subcritical process

In the subcritical case, the process becomes extinct with probability 1. What can be said about the asymptotic behavior?

Example: Linear Fractional Case. The probability of nonextinction is now equal to

$$1 - f_n(0) = m^n\left(\frac{1-q}{m^n - q}\right)$$

(cf. Section 3.1.4), which yields

$$\mathrm{E}(Z_n|Z_n > 0) = \frac{\mathrm{E}(Z_n)}{1 - f_n(0)} = \frac{m^n - q}{1-q} \to \frac{q}{q-1}, \quad n \to \infty.$$

This suggests that conditioning on nonextinction is a sufficient device to obtain a limit law. The proof of the following result can be found in the book of Athreya and Ney (1972).

Theorem 7 (Yaglom's). *If $m < 1$, then $\mathrm{P}\{Z_n = j|Z_n > 0\}$ converges, as $n \to \infty$, to a probability function whose generating function $\mathcal{B}(s)$ satisfies*

$$\mathcal{B}[f(s)] = m\mathcal{B}(s) + (1-m). \tag{3.21}$$

Also,

$$1 - f_n(0) \sim \frac{m^n}{\mathcal{B}'(1-)}, \quad n \to \infty. \tag{3.22}$$

The above theorem will be useful in the next application considered in this chapter. Convergence to a limit distribution conditional on nonabsorption is known as quasistationarity (see Sections 7.4 and 7.5).

3.5.3 *Critical process*

By analogy to the deterministic case, it might appear that in the critical case, in which $W_n = Z_n$, the sequence Z_n might reach a nontrivial limit. However, it is impossible, because the extinction probability is equal to 1 in this case. To approximate the asymptotic behavior of the critical Galton–Watson process, it is necessary to use conditioning on nonextinction and normalization. Let us start with a basic lemma (Athreya and Ney, 1972).

Lemma 3. *If* $m = \mathrm{E}(Z_1) = 1$ *and* $\sigma^2 = \mathrm{Var}(Z_1) < \infty$, *then*

$$\lim_{n\to\infty} \frac{1}{n}\left(\frac{1}{1-f_n(t)} - \frac{1}{1-t}\right) = \frac{\sigma^2}{2}$$

uniformly for $0 \le t < 1$.

Example: Linear Fractional Case. If $m = 1$, then $\mathrm{Var}(Z_1) = f''(1-) = 2p/(1-p) < \infty$, based on Eq. (3.10). In this case, the assertion of Lemma 3 is obtained by directly computing the limit (cf. Section 3.1.4).

Based on Lemma 3, the rate at which the critical process becomes extinct can be estimated. The limit behavior of the probability of nonextinction, $\mathrm{P}\{Z_n > 0\}$, is found by setting $t = 0$ in Lemma 3:

$$\mathrm{P}\{Z_n > 0\} = 1 - f_n(0) \sim \frac{2}{n\sigma^2}, \quad n \to \infty.$$

By the same token, the expectation of the process, given nonextinction, satisfies

$$\mathrm{E}(Z_n | Z_n > 0) = \frac{1}{\mathrm{P}\{Z_n > 0\}} \sim \frac{n\sigma^2}{2}, \quad n \to \infty.$$

This latter suggests that a limit law could exist for the normalized and conditional process $\{\frac{Z_n}{n} | Z_n > 0\}$. Indeed, we have the following:

Theorem 8. *If* $m = 1$ *and* $\sigma^2 < \infty$, *then*

$$\lim_{n\to\infty} \mathrm{P}\left\{\frac{Z_n}{n} > z | Z_n > 0\right\} = \exp\left(-\frac{2z}{\sigma^2}\right), \quad z \ge 0.$$

For the proof, see Athreya and Ney (1972).

3.6 *Application*: Gene Amplification

Material of this section is based on the article by Kimmel and Axelrod (1990). It is an example of application of the Yaglom's theorem (Theorem 7) to the analysis of the asymptotic behavior of a subcritical Galton–Watson process.

3.6.1 Gene amplification and drug resistance

Amplification of a gene is an increase of the number of copies of that gene in a cell. Amplification of genes coding for the enzyme dihydrofolate reductase (DHFR) has been associated with cellular resistance to the anticancer drug methotrexate (MTX).

A resistant population with an increased number of DHFR gene copies per cell can be obtained after a sensitive population is grown in increasing concentrations of the drug. Increased resistance is correlated with increased numbers of gene copies on small extrachromosomal DNA elements. These elements are visible in the microscope and resemble pairs of small chromosomes; they are called double minute chromosomes or double minutes. The number of DHFR genes on double minutes in a cell may increase or decrease at each cell division. This is because double minutes are acentric (i.e., they do not have centromeres like real chromosomes). Centromeres are required for the mitotic apparatus to faithfully segregate chromosomes into progeny cells.

In populations of cells with the double minutes, both the increased drug resistance and the increase in number of gene copies are reversible. The classical experiment confirming this includes transferring the resistant cell population into a drug-free medium. When these populations are grown in the absence of the drug, they gradually lose resistance to the drug, by losing extra gene copies.

The population distribution of numbers of copies per cell can be estimated by the experimental technique called flow cytometry. In the experiments described, two features of these distributions are notable. First, as expected, the proportion of cells with amplified genes decreases with time. Second, less obvious, the shape of the distribution of gene copy number within the subpopulation of cells with amplified genes appears stable as resistance is being lost. This stable distribution is depicted in Figure 3.5, taken from Brown et al. (1981). The distribution of cells with amplified genes retains its shape; only the area under the distribution gradually decreases while the peak corresponding to sensitive cells increases.

3.6.2 Galton–Watson process model of gene amplification and deamplification

We consider a cell, one of its progeny (randomly selected), one of the progeny of that progeny (randomly selected), and so forth. The cell of the nth generation contains Z_n double minutes carrying the DHFR genes. During cell's life, each double minute is either replicated, with probability a, or not replicated with probability $1-a$, independently of the other double minutes. Then, at the time of cell division, the double minutes are segregated to progeny cells. If the double minute has not been replicated, it is assigned to one of the progeny cells with probability $\frac{1}{2}$. If it has been replicated, then either both copies are assigned to progeny 1 (wp $\alpha/2$), or to progeny 2 (wp $\alpha/2$), or they are divided evenly between both progeny (wp $1-\alpha$). Let us note that the two double minutes segregate independently to progeny cells only when $\alpha = \frac{1}{2}$. Otherwise, they either preferentially go to the same cell

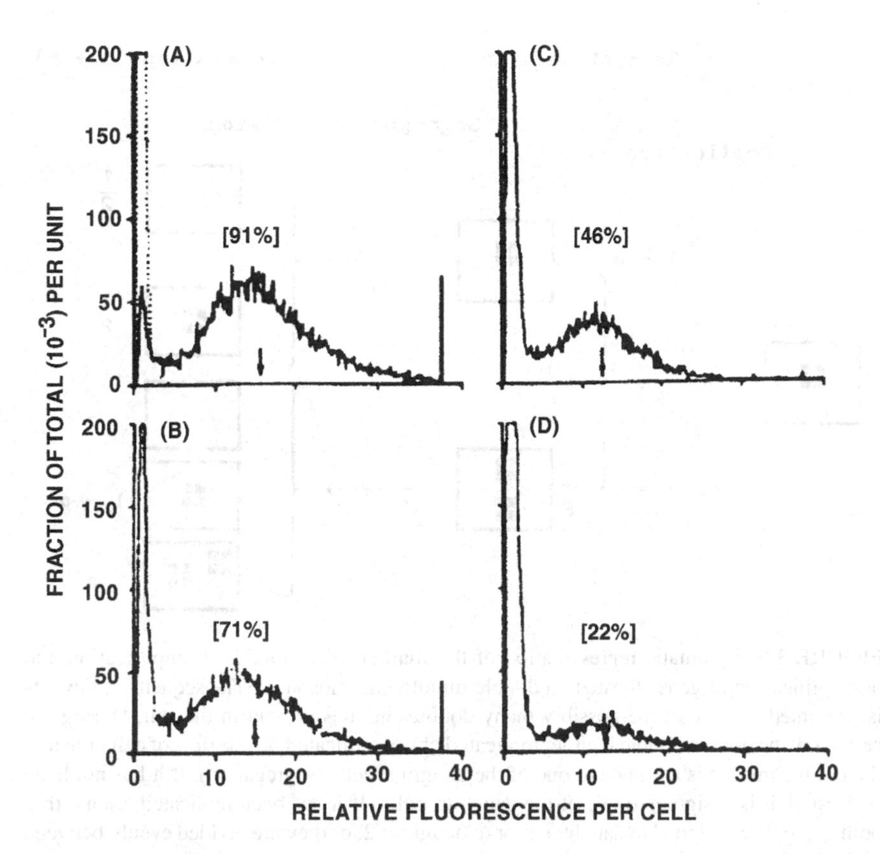

FIGURE 3.5. Loss of the amplified copies of the DHFR gene during cell growth in MTX-free media. The 3T6 cells resistant to the MTX were grown for different times in MTX-free medium. The fluorescence level is proportional to the number of gene copies per cell. The values in parentheses are the percentages of cells with gene copy numbers greater than those for sensitive cells. (A) Dotted line, 3T6 sensitive cells; solid line, resistant cells. (B) Cells grown for 17 generations without MTX. (C) Cells grown for 34 generations without MTX. (D) Cells grown for 47 generations without MTX. (Modified from Brown et al., 1981.) Source: Brown, P., Beverley, S.M. and Schimke. R.T. 1981. Relationship of amplified dihydrofolate reductase genes to double minute chromosomes in unstably resistant mouse fibroblast cell lines. Molecular and Cellular Biology 1: 1077–1983. Figure 1, page 1079. Copyright: 1981 American Society for Microbiology.

($\alpha > \frac{1}{2}$) or to different cells ($\alpha < \frac{1}{2}$). The randomly selected progeny in our line of descent contains one of the following (Fig. 3.6) the following:

- No replicas of the original double minute [wp $(1-a)/2 + a\alpha/2$]
- One replica of the original double minute [wp $(1-a)/2 + a(1-\alpha)$]
- Both replicas of the original double minute (wp $a\alpha/2$).

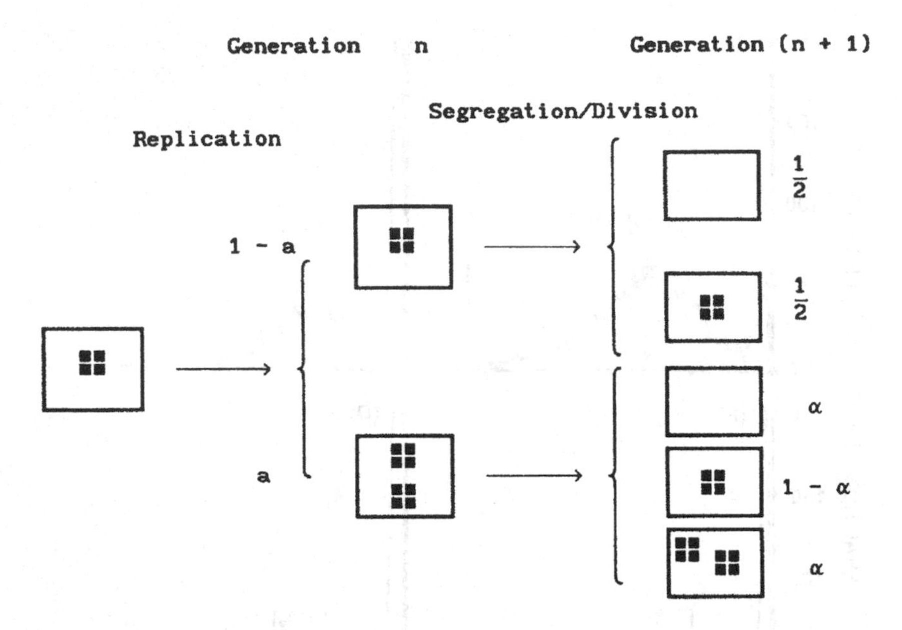

FIGURE 3.6. Schematic representation of the mathematical model of amplification and deamplification of genes located on double minute chromosomes. The sequence of events is presented for one of the possibly many double minutes present in the cell. During the cell's life, the double minute is either replicated or not replicated. At the time of cell division, the double minute is assigned to one of the daughter cells (segregation). If it has not been replicated, it is assigned to one of the daughter cells. If it has been replicated, then either both copies are assigned to daughter 1, or to daughter 2, or they are divided evenly between both daughters.

Therefore, the number of double minutes in the nth generation of the cell lineage is a Galton–Watson process with the progeny pgf:

$$f(s) = d + (1 - b - d)s + bs^2, \tag{3.23}$$

where $b = a\alpha/2$ and $d = (1 - a)/2 + a\alpha/2$ are the probabilities of gene amplification and deamplification, respectively. Because double minutes gradually disappear from the cell population in the absence of selection, it is assumed that deamplification (loss of gene copies) exceeds amplification, so that the process is subcritical. In mathematical terms, $b < d$ and $m = f'(1-) = 1 + b - d < 1$.

3.6.3 *Mathematical model of the loss of resistance*

We will call a cell resistant if it carries at least one double minute chromosome containing the DHFR gene. Otherwise, it is called sensitive. In the experiments described earlier, a population of cells resistant to MTX, previously cultured for N generations in medium containing MTX, consists only of cells with at least one DHFR gene copy (i.e., $Z_N > 0$). Therefore, the number of gene copies per cell is distributed as $\{Z_N|Z_N > 0\}$. If N is large, then, because the process is subcritical, by Theorem 7 (Yaglom's Theorem), this distribution has pgf $\mathcal{B}(s)$ satisfying the functional equation given in the theorem. Also, based on the estimates of $1 - f_n(0)$ provided in the same theorem, the resistant clone grows, in each generation, by the factor $2m$ on the average.

After the N initial generations, the resistant clone has been transferred to the MTX-free medium. The overall number of cells now grows by a factor of 2 in each generation, whereas the average number of resistant cells continues to grow by a factor of $2m$. Let us denote by $R(n)$ and $S(n)$ respectively the number of resistant and sensitive cells in the population, n generations after transferring the cells to the MTX-free medium; $r(n) = R(n)/[R(n) + S(n)]$ is the fraction of resistant cells. We obtain

$$R(n) = (2m)^n R(0), \quad S(n) + R(n) = 2^n[S(0) + R(0)];$$

hence,

$$\frac{r(n)}{r(0)} = m^n. \tag{3.24}$$

This means that the proportion of resistant cells decreases geometrically, whereas the distribution of gene copy number among the resistant cells remains close to the limit distribution of the Yaglom Theorem. This behavior is consistent with the experimental data of Figure 3.5.

3.6.4 *Probabilities of gene amplification and deamplification from MTX data*

Probabilities b and d can be estimated from the loss of resistance experiments similar to that in Kimmel and Axelrod (1990), using data on the S-$180(R_1A)$ cells in Kaufman et al. (1981). The resulting estimates are $b = 0.47$, and $d = 0.50$, yielding $a = 1 - 2(d - b) = 0.94$ and $\alpha = 2b/a = 1$. The interpretation is that although the frequency of replication of the double minute chromosomes is quite high, both copies are assigned almost always to the same progeny cell.

In Kimmel and Axelrod (1990), other models of the same process have been considered. All of them exhibit dynamics similar to that predicted by Yaglom's Theorem.

3.7 *Application*: Iterated Galton–Watson Process and Expansion of DNA Repeats

We consider mathematical properties of a time-discrete stochastic process describing explosive proliferation of DNA repeats in a class of human genetic diseases. The process contains copies of the Galton–Watson process as its building blocks.

3.7.1 *Dynamics of DNA repeats in human pedigrees*

Recently, several heritable disorders have been associated with dynamic increases of the number of repeats of DNA triplets in certain regions of the human genome. In two to three subsequent generations, the transitions from normal individuals to nonaffected or mildly affected carriers and then to full-blown disease occur. The two syndromes for which the most comprehensive data exist are as follows (Richards and Sutherland, 1994):

- The fragile X syndrome, caused by a mutation of the FMR-1 gene characterized by expansion of the $(CCG)_n$ repeats (normal 6–60; carrier 60–200, affected >200 repeats).
- Myotonic dystrophy, caused by a mutation of the DM-1 autosomal gene characterized by expansion of the $(AGC)_n$ repeats (normal 5–27, affected >50 repeats).

These two inherited human syndromes previously were distinguished by two features inconsistent with Mendelian inheritance: progressively earlier onset of symptoms in subsequent generations and higher severity of symptoms in subsequent generations.

These features have recently been correlated with changes in DNA. In each case, a trinucleotide existing in a few copies in an unaffected parent is found in multiple tandem copies in the progeny. The number of tandem copies is dramatically increased (10 – 100-fold) in affected individuals. It has been correlated with the age of onset and the severity of symptoms.

Important questions that have not been fully answered are as follows:

1. What is the mechanism of fluctuation of the number of repeat sequences in normal people (not in affected families)?
2. What is the mechanism of the modest increase in repeat sequences in unaffected carriers?
3. What is the mechanism of the rapid expansion of the number of repeat sequences in affected progeny within one or two generations?

Caskey et al. (1992) formulated a biological hypothesis regarding the origin of high variation in repeat count: "The instability in the premutation alleles which leads to the extraordinary expansions observed in DM and fragile X patients results from the presumed difficulty of replicating long GC-rich sequences. In this scenario, unequal rates of DNA synthesis lead to multiple incomplete strands of complementary, triplet, reinitiated sequences."

3.7.2 *Definition of the process*

Gawel and Kimmel (1996) make this hypothesis specific by assuming the following scenario of expansion of repeats:

- In the initial, 0th, replication round, the number of repeats is n.

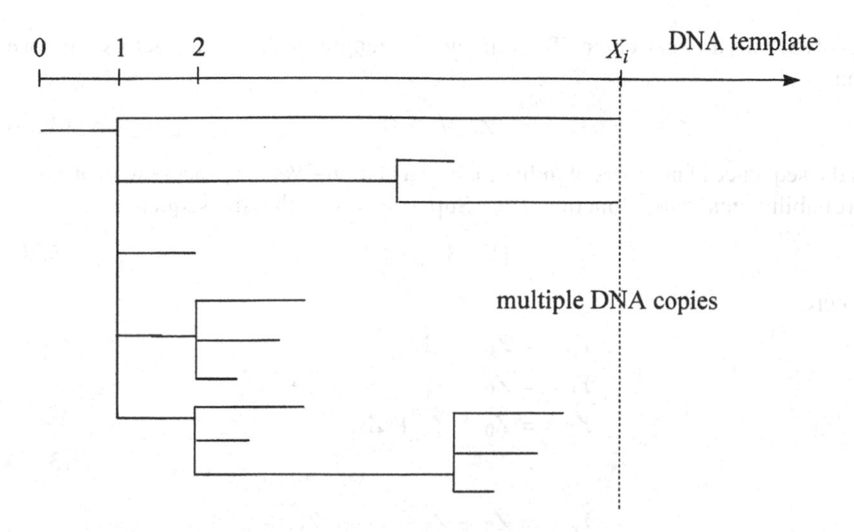

$$X_{i+1} = \Sigma\{\text{lengths of branches in a tree of height } X_i\} = Y_{X_i - 1}$$

FIGURE 3.7. The nonlinear mechanism of repeat expansion: Illustration of DNA branches that can be resolved into repeats. Suppose that there are X_i triplets in generation i and that a random number (usually 0 or 1) of new branches of DNA emerge on top of a previous one at the endpoint of each repeat (a single "initiation before termination" on each branch). Assuming that the process is confined to a region defined by the length of the original X_i repeats and that all triplets from all branches are resolved and incorporated into a linear structure of chromosomal DNA, we obtain the number of repeats X_{i+1} in the $(i + 1)$st generation. Source: Gawel, B. and Kimmel, M. 1996. The iterated Galton–Watson process. Journal of Applied Probability 33: 949–959. Figure 1, page 952. Copyright: 1996 Applied Probability Trust.

- In each new DNA replication round, a random number of new branching events (i.e., "initiation without termination of replication" events) occur at the endpoint of each repeat [this random number is characterized by the pgf $f(s)$].
- All resulting branches become resolved and reintegrated into the linear DNA structure, which becomes the template for the succeeding replication round.

Gawel and Kimmel (1996) note that there exists a precedent for such mechanism in the replication of the T4 bacteriophage. This virus induces production in the host cell of branched networks of concatenated DNA, which subsequently is resolved into unbranched phage genomes [see references in Gawel and Kimmel (1996)].

Gawel and Kimmel (1996) developed the so-called iterated simple branching process $\{X_i\}$ to provide a mathematical formulation for the expansion process (Figure 3.7). Here, X_i is the length of a linear chain of DNA repeats after the ith stage of replication ($i = 0, 1, 2, \ldots$) and $X_0 = n > 1$. A chain with $X_i = \nu$ repeats replicates as a branched network, which is assumed to be a Galton–Watson tree descended from a single ancestor through $\nu - 1$ generations. Thus, the replicating chain serves as a template for the height of the daughter tree. This partial tree later resolves into a linear chain. To compute the length of this chain, let us suppose that

$$\{Z_k,\ k \geq 0\} \tag{3.25}$$

is the sequence of numbers of individuals in a Galton–Watson process with progeny probability generating function $f(s)$. Suppose further that the sequence

$$\{Y_k,\ k \geq 0\}, \tag{3.26}$$

where

$$\begin{aligned} Y_0 &= Z_0 = 1, \\ Y_1 &= Z_0 + Z_1, \\ Y_2 &= Z_0 + Z_1 + Z_2, \\ &\vdots \\ Y_k &= Z_0 + Z_1 + \cdots + Z_k, \\ &\vdots \end{aligned} \tag{3.27}$$

is the total progeny process (i.e., Y_k is the cumulative number of progeny produced in the generations 0 through k of the Galton–Watson process) (cf. Section 3.10).

Further let, $\{Z_k^{(i)}, k \geq 0\}$, $i \geq 0$, be a sequence of iid copies of $\{Z_k\}$ with $\{Y_k^{(i)},\ k \geq 0\}$, $i \geq 0$ being the corresponding total progeny processes. These are the tree structures grown at each (ith) replication round. The generic process $\{Z_k\}$ is called the underlying Galton–Watson process.

The process

$$\{X_i, i \geq 0\}, \tag{3.28}$$

can be now defined in a recursive manner:

$$X_0 = n, \tag{3.29}$$

$$X_{i+1} = Y^{(i)}_{X_i - 1}, \quad i \geq 0. \tag{3.30}$$

Hence, the sequence $\{X_i\}$ is a Markov process and, because $Y_0^{(i)} = 1$, the state 1 is absorbing.

3.7.3 *Example*

The following version of the process seems to be realistic from the biological viewpoint. Suppose that at the end of each repeat, a new "initiation before termination" event occurs with small probability p, so that

$$f(s) = (1-p)s + ps^2. \tag{3.31}$$

Then, the number of branches stemming from each ramification point is at least one and at most two, the latter event being less likely. This leads to a "sparse" tree and implies that the growth of the process will be slow for a number of generations.

Fluctuations of the number of triplets in the unaffected individuals can be explained by coexistence of processes of triplet increase and triplet loss. Accordingly, we also assume that the process of resolution and reincorporation of repeats into the linear chromosomal structure has a limited efficiency $u < 1$.

This can be mathematically formalized using the idealized binomial thinning, (i.e., assuming that each repeat is resolved and reinserted with probability u). The new process $\{\tilde{X}_i, i \geq 0\}$ including the imperfect efficiency is defined as

$$\tilde{X}_0 = n, \tag{3.32}$$

$$\tilde{X}_{i+1} = B(u,\ Y^{(i)}_{\tilde{X}_i - 1} - 1) + 1, \quad i \geq 0, \tag{3.33}$$

where, conditional on N, $B(u, N)$ is a binomial random variable with parameters u and N.

With an appropriate choice of parameters (see Theorem 10 further on), this process produces runs of fluctuations, followed by explosive growth.

3.7.4 Properties

For the process without thinning, Pakes (2000) provides the following analysis, which is simpler than the original arguments in Gawel and Kimmel (1996). We exclude the trivial case $p_1 = 1$, where $X_i = X_0$. Then, $\mathrm{P}[\{X_i \to 1\} \cup \{X_i \to \infty\}] = 1$. Let X_∞ denote the almost sure limit of X_i and let $g(s, \nu)$ denote the pgf of Y_ν. Then, $g(s, 0) = s$ and $g(s, \nu + 1) = sf[g(s, \nu)]$ (see Section 3.10). It follows from Eq. (3.33) that

$$\mathrm{E}(s^{X_{i+1}}) = \mathrm{E}[g(s, X_i - 1)]$$

and, hence, in all cases,

$$\mathrm{E}(s^{X_\infty}) = \mathrm{E}[g(s, X_\infty - 1)]. \tag{3.34}$$

If

$$0 < p_0 < 1, \tag{3.35}$$

we may choose $s \in (0, q)$ and then $f(s) > s$. This gives $g(s, 1) > sf(s) > s$ and, hence, by induction, that $g(s, \nu - 1) > s^\nu$. Because Eq. (3.34) can be written as

$$s + \mathrm{E}(s^{X_\infty}, X_\infty > 1) = s + \mathrm{E}[g(s, X_\infty - 1), X_\infty > 1],$$

it is clear that this can hold if and only if $\mathrm{P}[X_\infty > 1] = 0$. We conclude that the process is absorbed at unity when Eq. (3.35) holds. Next, if $p_0 = 0$, then $Y^{(i)}_\nu > \nu + 1$ and, hence, Eq. (3.30) implies $X_{i+1} \geq X_i$. So, $X_i \uparrow \infty$ if $X_0 \geq 2$. The above reasoning (and some other details) are summarized by the following statement.

Theorem 9. *Let us consider the iterated Galton–Watson (IGW) process with no thinning (i.e., with $u = 1$). Then,*

1. $m < 1$ *yields* $\mathrm{E}(X_i) \to 1$ *and* $X_i \xrightarrow{a.s.} 1$;
2. $m = 1$ *yields* $\mathrm{E}(X_i) = \mathrm{E}(X_0)$ *and* $X_i \xrightarrow{a.s.} X_\infty$, *where* X_∞ *is a finite rv and* $X_\infty = 1$ *if* $p_0 < 1$;

3. *$m > 1$ yields* $\mathrm{E}(X_i) \to \infty$ *and*

a. *if* $p_0 > 0$, *then* $X_i \overset{a.s.}{\to} 1$,

b. *if* $p_0 = 0$ [*i.e.*, $f(s) = p_1 s + p_2 s^2 + \cdots$], *then* $X_i \overset{P}{\to} \infty$.

The next result concerns the growth of the IGW process with binomial thinning.

Theorem 10. *Suppose* $\{\tilde{X}_n\}$ *is the IGW process with binomial thinning.*

1. *Suppose* $m > 1$. *For each integer* $M > 0$, *there exists an integer* $N_0 > 0$ *such that*

$$\mathrm{E}(\tilde{X}_{i+1}|\tilde{X}_i = N_0) > MN_0.$$

2. *Suppose* $u > \frac{1}{2}$ *and* $p_0 = 0$. *There exist* $N_0 \geq 0$ *and* $\alpha > 1$ *such that*

$$\mathrm{E}(\tilde{X}_{n+1}|\tilde{X}_n \geq N_0) \geq \alpha \mathrm{E}(\tilde{X}_n - 1|\tilde{X}_n \geq N_0).$$

The properties stated in Theorem 9 are similar to those of the Galton–Watson process, with the absorbing state being $\{X = 1\}$ in our case, as opposed to $\{X = 0\}$ for the Galton–Watson process. The most notable difference is that the supercritical Galton–Watson process never becomes absorbed with probability 1, whereas the iterated supercritical process may.

Theorem 10 shows that no matter how small the efficiency u in the process with thinning, the process will increase (in the expected value sense) by an arbitrary factor, if only it exceeds a certain threshold. To illustrate the properties of the process with thinning, 20 independent simulations with parameters $p = 0.05$ and $u = 0.8$ are presented in Figure 3.8. All of them start from $n = 20$ repeats. Once the fluctuation exceeds 100–200 repeats, it usually jumps to ≥ 1000 repeats.

3.8 *Application*: Galton–Watson Processes in a Random Environment and Macroevolution

In evolutionary biology, it is frequently assumed that the environment of a population is fluctuating randomly (Gillespie 1986). If the dynamics of a population is described by a Galton–Watson branching process, this means that the pgf of the number of progeny per particle varies randomly from one generation to another.

The following account follows unpublished lecture notes by V. Vatutin (personal communication; also, consult Borovkov and Vatutin 1977). Assume that the reproduction law in a Galton–Watson process is changing from generation to generation and particles of the mth generation produce offspring according to the probability generating function $f_m(s)$. Clearly,

$$F_n(s) = F_{n-1}[f_n(s)] = f_0(f_1(\cdots(f_n(s))\cdots))$$

is the probability generating function specifying the distribution law of Z_n. One important case is the randomly changing environment. Specifically, let us define a collection $\mathcal{G} = \{G_a : a \in \mathcal{A}\}$ of probability generating functions with $\mathcal{A}$ being

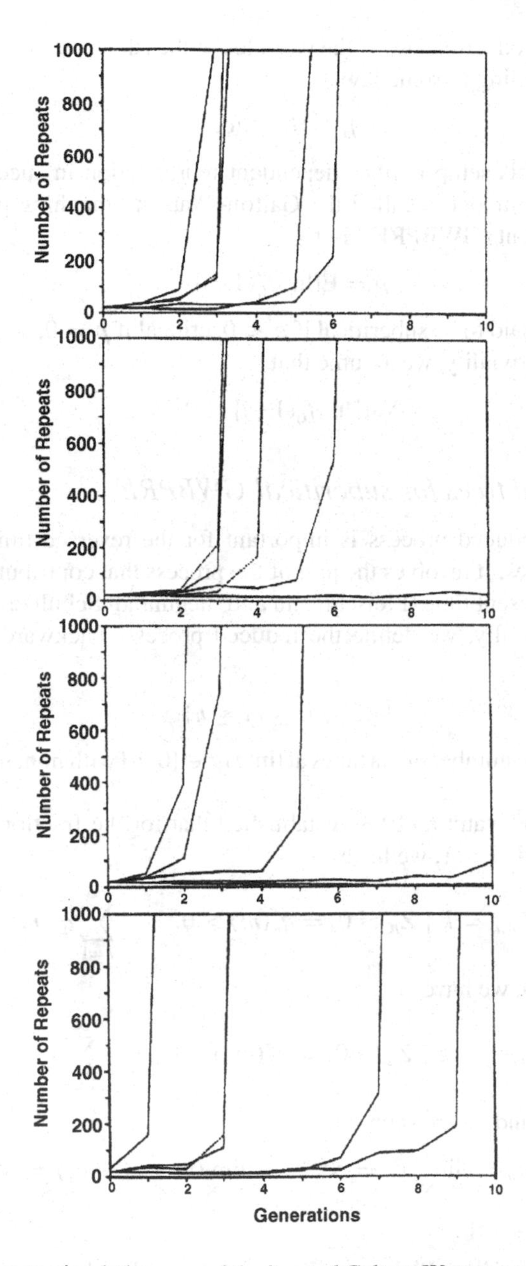

FIGURE 3.8. Twenty simulation runs of the iterated Galton–Watson process with binomial thinning. Parameters are $p = 0.05$ and $u = 0.8$; that is, a new "initiation before termination" event occurs with probability 5% and the efficiency of the resolution and reincorporation process is 80%. Each run starts from exactly 20 repeats and continues to fluctuate within narrow limits for a variable number of generations. Once the fluctuation exceeds 100–200 repeats, it usually jumps to ≥ 1000 repeats. Source: Gawel, B. and Kimmel, M. 1996. The iterated Galton–Watson process. Journal of Applied Probability 33: 949–959. Figure 2, page 958. Copyright: 1996 Applied Probability Trust.

some set. The reproduction law of the particles of the ith generation is taken from $\mathcal{G}$ at random according to some law

$$f_i \in \mathcal{G}, \quad \text{iid}.$$

Let us note that this setup implies dependent reproduction in successive generations. The above model is called the Galton–Watson branching processes in a random environment (GWBPRE). Let

$$\rho = \mathrm{E}[\ln\ f_0'(1-)].$$

The GWBPRE is said to be subcritical if $\rho < 0$, critical if $\rho = 0$, and supercritical if $\rho > 0$. For nontriviality, we assume that

$$\mathrm{Var}[\ln\ f_0'(1-)] > 0.$$

3.8.1 *Reduced trees for subcritical GWBPRE*

The concept of reduced process is important for the reversed-time analysis of branching processes. It involves the part of the process that contributed to individuals seen in the present time (Fleischmann and Siegmund–Schultze 1977, Sagitov 1989). Mathematically, we define the reduced process (backward genealogical tree) as a family

$$\{Z_{m,n},\ 0 \le m \le n\}$$

in which $Z_{m,n}$ is the number of particles at time $m \in [0, n]$ with nonempty offspring at time n.

Fleischmann and Vatutin (1999) established that for the fractional linear case (Section 3.1.4) and $m > 1$, we have

$$\lim_{n\to\infty} \mathrm{P}[Z_{m,n} = k \mid Z_n > 0] = q_k(m) > 0, \qquad \sum_{k=1}^{\infty} q_k(m) = 1,$$

and for all $m^* > 0$, we have

$$\lim_{n\to\infty} \mathrm{P}[Z_{n-m^*,n} = k \mid Z_n > 0] = q_k^*(m^*) > 0, \qquad \sum_{k=1}^{\infty} q_k^*(m^*) = 1,$$

and, finally, if u_n and v_n are such that

$$\lim_{n\to\infty} u_n = \lim_{n\to\infty} v_n = \infty, \qquad \lim_{n\to\infty} (n - u_n - v_n) = \infty,$$

then

$$\lim_{n\to\infty} \mathrm{P}[Z_{u_n,n} = Z_{n-v_n,n} \mid Z_n > 0] = 1. \tag{3.36}$$

Let us assume that the present time is n, in the units of one generation of particles. If we observe a nonextinct process population that evolved in the past like a subcritical GWBPRE, we see that with a high probability, during the long time interval $[u_n, n - v_n]$, the process did not change state. This means that the divergence happened either very close to the present moment or very far in the past.

3.8.2 *Evolutionary interpretation*

V. Vatutin (personal communication) noted that Eq. (3.36) may enable a reinterpretation of conclusions based on molecular evidence of genetic divergence between humans and chimpanzees. One of the more influential recent evolutionary theories is the theory of punctuated equilibria. The theory, based on some fossil evidence, states that long periods of evolutionary stasis (invariance of species) are interspersed with bursts of speciation (appearance of new species). If the evolutionary process can be modeled using a subcritical GWBPRE, then the observed periods of evolutionary stasis preceded and followed by bursts of speciation may not necessarily reflect the unevenness of the evolutionary process itself, but they may follow from the properties of the reduced GWBPRE. Gillespie's (1986) more general observations concerning the evolution's "episodic clock" can be similarly re-interpreted. Gillespie (1986) has investigated the number of substitutions in DNA and protein. He found the ratio of the variance to the mean in a set of four nuclear and five mitochondrial genes in mammals ranged from 0.16 to 35.55, which can be interpreted as periods of stasis alternating with periods of rapid substitution. To fit these data, Gillespie (1986) suggested models that incorporate natural selection in a changing environment. Reduced GWBPRE might provide an alternative for these models.

3.9 Other Works and Applications

Much work was carried out concerning both various generalizations of the Galton–Watson process and diverse properties of the basic process. Further in this book, we will consider examples of Galton–Watson processes with diverse type spaces. In this section, we provide examples of a different kind.

3.9.1 *Stochastic dependence*

Stochastic dependence in branching processes can be formulated in various ways. Examples include intergeneration dependence and dependence between relatives. Both are interesting because of their applications in cell proliferation. It is known that progeny cells emerging from a division of a parent cell have life lengths and other parameters which are correlated. A number of researchers attempted to account for these empirical observations (Axelrod and Kuczek 1989, Brooks et al. 1980, Rigney 1981, Hejblum et al. 1988, Kuczek and Axelrod 1986, Sennerstam and Strömberg 1996, Staudte et al. 1984, 1997, Webb, 1989).

Generation dependence (Fearn 1972) has a different meaning for the Galton–Watson process in which the generations are synchronized, and in the time-continuous age-dependent processes (Fearn 1976), in which the generations overlap (Chapters 4 and 5). One way of capturing dependence between relatives is to consider the individual together with his/her relatives (siblings, cousins, etc.) as a single superindividual. This can be carried out using the framework of general processes (Olofsson 1996). For a different approach, see Crump and Mode (1969). In the framework of estimation, a convenient manner of expressing such "local" dependencies is the bifurcating autoregression (Section 5.5.3).

3.9.2 *Process state dependence*

All Galton–Watson processes, including these for which the progeny distributions depend on the state of the process, are Markov. However, there is no simple relationship linking the type of dependence with the properties of the resulting Markov chain. Therefore, the study of such processes proceeds by way of special cases, deemed important usually for extramathematical reasons. An early reference is the article by Lipow (1975).

A series of articles by Klebaner consider limit properties of processes with progeny distributions depending on the process state (i.e., usually the number of particles at a given time). Klebaner (1997) provided a short review of size- and density-dependent processes. Klebaner (1988, 1990) and Cohn and Klebaner (1986) consider applications in demography and genetics. Another interesting article (Klebaner and Zeitouni 1994) considers the problem of "cycle slip" (i.e., the conditions that a randomly perturbed deterministic system has to satisfy to escape the basin of attraction of the deterministic part).

Another application is presented by Jagers (1995), who used the coupling method to analyze state-dependent processes describing proliferation of biological cells.

3.9.3 *Bisexual Galton–Watson process*

The bisexual generalization of the Galton–Watson process is not straightforward to consider, because it involves a process of pair formation. One way to proceed is to assume that only females bear progeny, of both genders, and to define a mating function which provides, for each unpaired female, the probability of forming a pair and mating with an available male. These functions may be consistent with monogamy or monoandry or they may mimic the mating patterns of insects and so forth. The mating process destroys the branching property and the resulting stochastic process is not strictly speaking a branching process. One of the recent articles on the limit properties of such processes is by González and Molina (1996). An exhaustive review of older and current literature is provided in the thesis by Falahati (1999).

3.9.4 *Age of the process*

The estimation of the age of the branching process based on data concerning extant individuals, their number, types, and so forth gained importance because of applications in genetics and molecular evolution. The evolution of chromosomes containing disease genes can be represented as a branching process with a Poisson distribution of progeny if the disease subpopulation is a small subset of a larger population evolving according to the Fisher–Wright model. A model of this type was considered by Kaplan et al. (1995) and used to obtain simulation-based likelihood estimates of location and age of disease genes. A number of refinements of this method can be found in the unpublished doctoral thesis of Pankratz (1998), in which further references also are provided.

Another type of application is finding the age of the most recent common ancestor of a population characterized by its genetic makeup, under the assumption that its demography followed a branching process. An example related to the evolution of modern humans, using a time-continuous branching process, is described in detail in Section 4.4.

An early article concerning the estimation of the age of a Galton–Watson branching process is by Stigler (1970). The author uses the fractional linear case, in which an estimator can be explicitly derived, and then generalizes the results to the case of the general Galton–Watson process. This article was followed by a number of other publications, including those by Tavaré (1980) and Koteeswaran (1989).

3.9.5 *Family trees and subtrees*

A somewhat related subject is the probability that the family tree of the process contains an infinite N-nary subtree (i.e., a tree with exactly N progeny of each individual). Pakes and Dekking (1991) demonstrated that this probability is the largest root in the interval [0, 1] of the equation

$$1 - t = G_N(1 - t),$$

where

$$G_N(s) = \sum_{j=0}^{N-1} \frac{(1-s)^j f^{(j)}(s)}{j!}$$

and $f(s)$ is the offspring distribution of the process. Further results concerning the maximum height of the N-nary subtree are provided in the same article.

3.10 Problems

1. Following are given several examples of probability generating functions of a Galton–Watson process. For each of them, find $\mathrm{E}(Z_1) \equiv m$ and $\mathrm{Var}(Z_1) \equiv \sigma^2$. Assume that the Galton–Watson process describes a cell population with discrete generations. Characterize the model described by each pgf. *Example.* If $f(s) = (ps + q)^2$, then each of the two daughter cells, independently, survives with probability p and dies with probability q.
   - $f(s) = ps^2 + qs$
   - $f(s) = ps^2 + q$
   - $f(s) = ps^2 + qs + r$

2. Assume that the pgf of the Galton–Watson process is the fractional linear function. Using induction, prove the form of $f_n(s)$ in the case $m = 1$.
3. Assume the fractional linear case. Treating the Galton–Watson process as a Markov chain, check that the state $\{Z_n = k\}$ is transient if $k \neq 0$ and recurrent if $k = 0$. *Hint:* Use the closed form of $f_n(s)$ and base the assertion on the condition of divergence of $\sum_{n\geq 0} \Pr\{Z_n = k\}$.
4. Suppose that a Galton–Watson process with the pgf $f(s)$ is started not by a single particle, but by a random number of particles [with pgf $g(s)$]. Find $f_n(s)$.
5. *Problem 4 Continued.* Assume $f(s)$ the fractional linear function with $m < 1$ and $g(s) = (q-1)s/(q-s)$, where $q = f(q)$. Define $\bar{f}_{(n)}(s) = [f_n(s) - f_n(0)]/[1 - f_n(0)]$, the conditional pgf of Z_n provided $Z_n > 0$. Prove, using induction, that $\bar{f}_{(n)}(s) \equiv g(s)$.
6. Distribution with pgf $g(s)$ having properties as in the previous problem is called a quasistationary distribution of the Galton–Watson process. What makes it different compared to the stationary distribution of a Markov chain?
7. *Galton–Watson Process in Varying Environment.* Suppose that the nth generation of particles has the progeny distribution $\{p_k^{[n]},\ k \geq 0\}$ with pgf $f^{[n]}(s)$. Define the process in the terms of a Markov chain and derive the forward equation as it was done for the ordinary process. What is $f_n(s)$ now?
8. *Integrated Galton–Watson Process* . Consider the process $\{Y_n\}$, where $Y_n = \sum_{i=0}^{n} Z_i$. Demonstrate that the pgf of Y_n, denoted $F_n(s)$, satisfies

$$F_{n+1}(s) = sf[F_n(s)].$$

9. *Problem 8 Continued.* Demonstrate that if $m < 1$, then the limit $\lim_{n\to\infty} F_n(s) = F(s)$ exists and satisfies the following functional equation:

$$F(s) = sf[F(s)].$$

Hint: Show that $|F_{n+1}(s) - F_n(s)| \leq m|F_n(s) - F_{n-1}(s)|$ if $s \in [0, 1]$. $F(s)$ is the pgf of the total number of particles produced in the process, and in the subcritical case it is a proper random variable [i.e., $F(1) = 1$].
10. *Problem 8 Continued.* Assume the linear fractional case and calculate $F(s)$ by solving the functional equation in Problem 9. Does $F(s)$ correspond to any standard discrete distribution?
11. *Quasistationary Distribution.* Suppose that a subcritical Galton–Watson (GW) process with the pgf $f(s)$ is started not by a single particle, but by a random number of particles having pgf $\mathcal{B}(s)$, defined in Yaglom's Theorem. Prove that this distribution is a stationary distribution of the GW process. *Hint:* Use the functional equation defining $\mathcal{B}(s)$ and the property that $\mathcal{B}(0) = 0$.
12. Assume the fractional linear case and $m > 1$. Calculate the Laplace transform of $W_n = Z_n/m^n$ and find its limit as $n \to \infty$. What is the distribution of W?
13. Consider the following mechanisms of gene amplification:
    - Each of the double minute chromosomes present in the newborn daughter cell survives wp p. If it does survive, then during replication, each next copy of this particular double minute chromosome is produced wp p.

- During segregation, each copy is assigned to given daughter cell wp $\frac{1}{2}$.

Consider a random lineage of cells in the population. If in the 0th generation there exists only a single cell with a single double minute chromosome, then $\{Z_n, n \geq 0\}$, the sequence of number of copies of the double minute in the cell of nth generation, forms a Galton–Watson process with the progeny pgf $f(s)$. Find $f(s)$. *Hint:* Use the expression for the pgf of the sum of random number of iid rv's.

14. *Problem 13 Continued.* Using the properties of the linear fractional pgf's, assuming the subcritical process, find the pgf $\mathcal{B}(s)$ of the limit distribution of the number of double minute chromosomes per cell in the cells of the resistant clone. Suppose that the mean number of double minute chromosomes per resistant cell is equal to 20 and that double minutes have been counted in 50 cells. Find the maximum likelihood estimate of p and an approximate 95% confidence interval for this estimate.

15. Consider a population of particles with life lengths equal to 1, proliferating by binary fission, with each of the two progeny surviving independently with probability p.

a. Find the probability of eventual extinction

$$q = \Pr\{\#\text{ particles } = 0 \text{ at some time } n\}$$

for a population started by a single ancestor particle, as the function of p [i.e., $q = q(p)$], for $p \in [0, 1]$.

b. Find the probability that at time $n = 3$, there will be four or less particles in the process.

c. An *ad hoc* way to increase the probability of nonextinction of the process is to start at time 0 from a collection of N ancestor particles, instead of 1. Find the probability $q = q(p, N)$ of eventual extinction of such process. For $p = \frac{3}{4}$, what should N be equal to so that $1 - q(p, N)$ exceeds 0.999?

16. Consider a Galton–Watson process Z_n with progeny pgf $h(s)$, started by a random number Y of ancestors [where $Y \sim g(s)$]. Find

a. $\mathrm{E}\,(Z_n|Z_0 = Y)$

b. $\mathrm{Var}(Z_n|Z_0 = Y)$

c. $\Pr\{Z_n = 0, \text{ some } n \;|Z_0 = Y\}$.

CHAPTER 4

The Age-Dependent Process: The Markov Case

This chapter is devoted to the use of the time-continuous branching process with exponential life time distributions. This process also has the Markov property and is closely related to the Galton–Watson process. The exponential distribution to model lifetimes of particles is not well motivated by any biological assumptions. Indeed, the exponential distribution admits lifetimes which are arbitrarily close to 0, whereas it is known that life cycles of organisms and cells have lower bounds of durations, which are greater than 0. The advantage of using the exponential distribution is that it leads, in many cases, to computable expressions. These latter allow one to deduce properties which then can be conjectured for more general models.

4.1 Differential Equation for the pgf and Its Elementary Properties

4.1.1 *Definition of the process*

The process can be described as follows. A single ancestor particle is born at $t = 0$. It lives for time τ, which is exponentially distributed with parameter λ. At the moment of death, the particle produces a random number of progeny according to a probability distribution with pgf $f(s)$. Each of the first-generation progeny behaves, independently of each other, in the same way as the initial particle. It lives for an exponentially distributed time and produces a random number of progeny. Progeny of each of the subsequent generations behave in the same way. If we denote the particle count at time t by $Z(t)$, we obtain a stochastic process $\{Z(t),\ t \geq 0\}$.

The probability generating function $F(s, t)$ of $Z(t)$ satisfies an ordinary differential equation which is easiest to derive based on the Markov nature of the process.

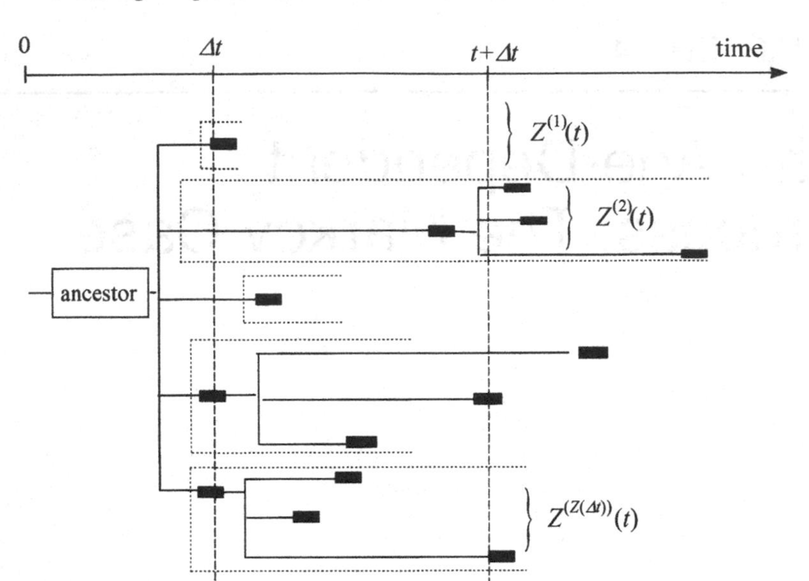

FIGURE 4.1. Derivation of the backward equation for the Markov time-continuous branching process.

Indeed, let us consider the process at a given time t. Any of the particles existing at this time, whatever its age is, has a remaining lifetime distributed exponentially with parameter λ. This follows from the lack of memory of the exponential distribution. Therefore, each of the particles starts, independently, a subprocess identically distributed with the entire process (Figure 4.1). Consequently, at any time $t + \Delta t$, the number of particles in the process is equal to the sum of the number of particles in all iid subprocesses started by particles existing at time Δt. Each of these subprocesses is of age t. In mathematical terms,

$$Z(t + \Delta t) = \sum_{i=1}^{Z(\Delta t)} Z^{(i)}(t), \tag{4.1}$$

where the superscript (i) identifies the ith iid subprocess. So, according to the pgf theorem (Theorem 1), we have the following pgf identity:

$$F(s,\ t + \Delta t) = F[F(s,\ t),\ \Delta t]. \tag{4.2}$$

We subtract $F(s, t)$ from both sides and, remembering that the process is started by a single particle [i.e., $F(s, 0) = s$], we can write the result in the following form:

$$F(s, t + \Delta t) - F(s, t) = F[F(s, t), \Delta t] - F[F(s, t), 0]. \tag{4.3}$$

If Δt is small, then with a probability close to 1, the process consists of only either the ancestor or of its first-generation progeny. In the terms of the process pgf,

$$F(s, \Delta t) = se^{-\lambda \Delta t} + f(s)(1 - e^{-\lambda \Delta t}) + o(\Delta t) \tag{4.4}$$

or

$$F(s, \Delta t) - F(s, 0) = [-s + f(s)](1 - e^{-\lambda \Delta t}) + o(\Delta t). \tag{4.5}$$

Substituting Eq. (4.5) into Eq. (4.3) and dividing by Δt, we obtain

$$\frac{F(s, t + \Delta t) - F(s, t)}{\Delta t} = \frac{\{-F(s, t) + f[F(s, t)]\}(1 - e^{-\lambda \Delta t}) + o(\Delta t)}{\Delta t}.$$

By letting $\Delta t \to 0$, this leads to the following differential equation:

$$\frac{dF(s, t)}{dt} = -\lambda\{F(s, t) - f[F(s, t)]\}. \tag{4.6}$$

Equation (4.6), with the initial condition $F(s, 0) = s$, has a unique pgf solution if conditions are satisfied which guarantee that the process does not explode [i.e., that at each time $t > 0$, the number of particles is finite wp 1 or that $\lim_{s \uparrow 1} F(s; t) = 1$ (Pakes 1993)]. For this, it is sufficient that the expected number of progeny per particle $m = f'(1-)$ be finite (Athreya and Ney 1972).

In particular, expression (4.2) demonstrates that for any time increment Δt, we have

$$F(s, i\Delta t) = f_{\Delta t}^{(i)}(s),$$

where $f_{\Delta t}^{(i)}(s)$ is the ith iterate of $F(s, \Delta t)$. Therefore, $\{Z(i\Delta t),\ i = 0, 1, \ldots\}$ is a Galton–Watson process with progeny pgf $f_{\Delta t}(s)$. Of course, $f_{\Delta t}(s)$ has properties very different from those of $f(s)$. In particular, even if $f(s)$ admits only a finite number of progeny, the distribution of $Z(i\Delta t)$ always has an infinitely long right tail.

4.1.2 Probability of extinction and moments

The Markov branching process is called

- *subcritical* if $m < 1$,
- *critical* if $m = 1$,
- *supercritical* if $m > 1$.

Let q be defined as for the Galton–Watson process {i.e., as the smallest root of the equation $f(s) = s$, $s \in [0, 1]$}. The extinction probability is, again, equal to q.

Theorem 11. *Suppose that* $m < \infty$. *If* $F(s; t)$ *is the pgf solution of Eq. (4.6), then* $P(t) \equiv F(0; t) \to q$ *as* $t \to \infty$.

The extinction probability result is the same as for the Galton–Watson process. The expressions for the moments are almost as simple as they are for the Galton–Watson process.

Let us define the kth factorial moment *of* $Z(t)$, $m_k(t) = \mathrm{E}\{Z(t)[Z(t) - 1] \cdots [Z(t) - k + 1]\}$. The differential equations for the factorial moments of the process are formally derived by differentiating Eq. (4.6) with respect to s and letting $s \uparrow 1$. For example, the expected value $m_1(t)$ satisfies

$$\frac{dm_1(t)}{dt} = \lambda(m - 1)m_1(t), \quad m_1(0) = 1.$$

These equations can be solved explicitly. We obtain the following expressions for the expectation and variance of $Z(t)$:

$$\mathrm{E}[Z(t)] = e^{at}, \tag{4.7}$$

$$\mathrm{Var}[Z(t)] = \begin{cases} \dfrac{f''(1-) - f'(1-) + 1}{f'(1-) - 1} e^{at}(e^{at} - 1), & a \neq 0 \\ f''(1-)\lambda t, & a = 0, \end{cases} \tag{4.8}$$

where $a = \lambda(f'(1-) - 1)$ is the Malthusian parameter of population growth.

4.2 *Application*: Clonal Resistance Theory of Cancer Cells

The aim of cancer chemotherapy is to achieve remission (i.e., the disappearance of clinically detectable cancers) and then to prevent relapse (i.e., the regrowth of cancer). In many cases, the failure of chemotherapy is associated with the growth of cells resistant to further treatment with the same drug. There are two conceivable modes of drug resistance: Resistant cells might exist in tumors before treatment and be selected for during treatment. Alternatively, they might be induced by treatment.

Drug resistance was extensively studied in bacteria (see Section 6.1 and also a review article by Levy 1998) and the resulting ideas have been applied to understanding drug resistance in cancer cells. One possible hypothesis is that mutations from sensitivity to resistance are rare, irreversible events that spontaneously occur in the absence of the selecting drug. Moreover, mutation to resistance to a drug is a single event and it arises independently of resistance to another drug. Although simplistic, this model is useful in understanding the initiation and growth of drug-resistant cancer cells. Also, it might help design new protocols of cancer chemotherapy.

We explore the branching process approach to a theory of resistance, which has become influential in the cancer research community. It was originally developed by Goldie and Coldman (1979, 1984), Goldie (1982) and Coldman and Goldie (1985). We will rederive some of the original results, using Markov time-continuous branching processes. This approach seems more rigorous.

The assumptions of the theory are as follows (Fig. 4.2).

1. The cancer cell population is initiated by a single cell which is sensitive to the cytotoxic (chemotherapeutic) agent. The population proliferates without losses.
2. The interdivision time of cells is a random variable with a given distribution.
3. At each division, with given probability, a single progeny cell mutates and becomes resistant to the cytotoxic agent.
4. Mutations are irreversible.

We wish to compute the probability that when the tumor is discovered, it does not contain any resistant cells. Only in such a situation, is the use of a cytotoxic

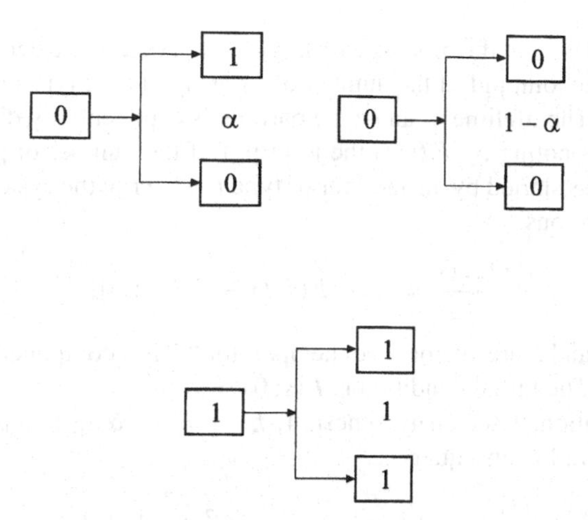

FIGURE 4.2. Schematic representation of the branching process of clonal resistance in the single-mutation case.

agent effective. If even a small subpopulation of resistant cells exists, the cancer cell population will eventually reemerge despite the therapy.

4.2.1 *Single-mutation case*

The branching process model

We translate the hypotheses of clonal resistance into the language of branching processes.

1. In the process, there exist two types of particle, labeled 0 (sensitive) and 1 (resistant).
2. The process is initiated by a single type 0 particle.
3. The life spans of particles are independent random variables, distributed exponentially with parameter λ.
4. Each particle, at death, divides into exactly two progeny particles:
   - The 0-particle produces either two 0-particles, wp $1-\alpha$, or one 0– particle and one 1-particle, wp α.
   - The 1-particle produces two 1-particles.

Thus, we have a two-type time-continuous Markov branching process.

Let us introduce the following notations, which are required because we consider two types of particle:

- $F_0(s_0, s_1; t)$ is the joint probability generating function (see Appendix A) of the numbers of cells of both types, present at time t in the process initiated at time 0 by a type-0 cell.
- $F_1(s_1; t)$ is the pgf of the numbers of cells of type 1, present at time t in the process initiated at time 0 by a type-1 cell.

Frequently, we will write $F_i(s; t)$ and even $F_i(t)$ or $F_i(s)$ or F_i.

In the general case of the process with k types of particles, we denote by $f_i(s) = f_i(s_1, \ldots, s_k)$ the joint pgf of the number of progeny of all k types begotten by an i-type particle. The lifetime of an i-type particle is exponentially distributed with parameter λ_i. Denoting by $F_i(s;t)$ the joint pgf of the number of particles of all types in a process started by an ancestor of type i, we write the system of ordinary differential equations

$$\frac{dF(s;t)}{dt} = -\lambda \cdot \{F(s;t) - f[F(s;t)]\}, \tag{4.9}$$

in which F, f, and λ are vectors and the operator "$\cdot$" is a componentwise product of two vectors. The initial condition is $F(s;0) = s$.

In our application, based on hypothesis 4, $f_0(s) = (1-\alpha)s_0^2 + \alpha s_0 s_1$, $f_1(s) = s_1^2$, and $\lambda_0 = \lambda_1 = \lambda$. In consequence,

$$\frac{dF_0}{dt} = -\lambda F_0 + \lambda[(1-\alpha)F_0^2 + \alpha F_0 F_1], \tag{4.10}$$

$$\frac{dF_1}{dt} = -\lambda F_1 + \lambda F_1^2. \tag{4.11}$$

Solutions

Finding explicit solutions for cell proliferation models of the type (4.10), (4.11) frequently leads to differential equations with right-hand sides quadratic in the unknown function (so-called Riccatti-type equations). The reason is that in such models, the pgf of the number of progeny is a second-order polynomial, which reflects the binary fission mode of proliferation of living cells. The following result can be verified by direct substitution. Uniqueness follows by the usual regularity conditions.

Theorem 12. *The solution of the differential equation*

$$\frac{dF(t)}{dt} = f(t)F(t) + hF(t)^2, \tag{4.12}$$

where $f \in C[0,\infty)$, with initial condition $F(0)$, is a uniquely defined function $F \in C^1[0,\infty)$

$$F(t) = \frac{F(0)e^{\int_0^t f(u)\,du}}{1 - hF(0)\int_0^t e^{\int_0^u f(v)\,dv}du}. \tag{4.13}$$

We will solve the system (4.10), (4.11). First, the separation of variables, or Eq. (4.13) is applied to Eq. (4.11) and it yields

$$F_1(s;t) = \frac{s_1}{s_1 + (1-s_1)e^{\lambda t}}. \tag{4.14}$$

Substituting Eq. (4.14) into Eq. (4.10) and employing Theorem 12, we obtain

$$F_0(s;t) = \frac{s_0 e^{-\lambda t}[e^{-\lambda t}s_1 + (1-s_1)]^{-\alpha}}{1 + s_0\{[e^{-\lambda t}s_1 + (1-s_1)]^{1-\alpha} - 1\}s_1^{-1}}. \tag{4.15}$$

Differentiating $F_0(s;t)$ with respect to s_0 and s_1, we obtain the expressions for the expected counts of the sensitive and resistant cells

$$M_0(t) = \frac{\partial F(1,1;t)}{\partial s_0} = e^{\lambda(1-\alpha)t}, \quad t \geq 0,$$
$$M_1(t) = \frac{\partial F(1,1;t)}{\partial s_1} = e^{\lambda t} - e^{\lambda(1-\alpha)t}, \quad t \geq 0.$$

The conclusion is that in the absence of intervention, the resistant cells eventually outgrow the sensitive ones. The probability of no resistant cells at time t is also easy to obtain:

$$P(t) = \lim_{s_0 \uparrow 1} \lim_{s_1 \downarrow 0} F_0(s;t) = \frac{1}{(1-\alpha) + \alpha e^{\lambda t}} = \frac{1}{(1-\alpha) + \alpha[M_0(t) + M_1(t)]}. \tag{4.16}$$

Conclusions

Based on Eq. (4.16), the following observations can be made (Coldman 1987, Coldman and Goldie 1983, 1985, 1987, Coldman et al. 1985)

- The probability that there are no resistant cells at time t is inversely related to the total number of cells.
- For different mutation rates α, if the α's are small, the plots of $P(t)$ are approximately shifted, with respect to each other, along the t axis.
- The time interval in which the resistant clone is likely to emerge [i.e., in which $P(t)$ falls from near 1 to near 0, e.g., from 0.95 to 0.05], constitutes a relatively short "window" (Fig. 4.3). Therefore, the therapy should be prompt and radical to decrease the cell number and probability $[(1-P(t))]$ of emerging resistance. For discussions, see e.g. Mackillop (1986) and Rosen (1986).

An alternative model

An alternative variant of the model presented above assumes that each of the progeny cells may mutate independently with probability α, as depicted in Figure 4.4.

The equations of the process assume now the form

$$\frac{dF_0}{dt} = -\lambda F_0 + \lambda[(1-\alpha)F_0 + \alpha F_1]^2, \tag{4.17}$$
$$\frac{dF_1}{dt} = -\lambda F_1 + \lambda F_1^2. \tag{4.18}$$

They are of a more general Riccati form, not admitting a closed-form solution. However, it is still possible to obtain $P(t)$. Let us note that $F_1(1,0;t) \equiv 0$ (i.e., the probability of no resistant cells in the subprocess initiated by a resistant cell is equal to 0). Therefore, letting $s_0 \uparrow 1$ and $s_1 \downarrow 0$ in Eq. (4.17) yields the following differential equation for $P(t)$:

$$\frac{dP(t)}{dt} = -\lambda P(t) + \lambda(1-\alpha)^2 P(t)^2, \quad P(0) = 1, \tag{4.19}$$

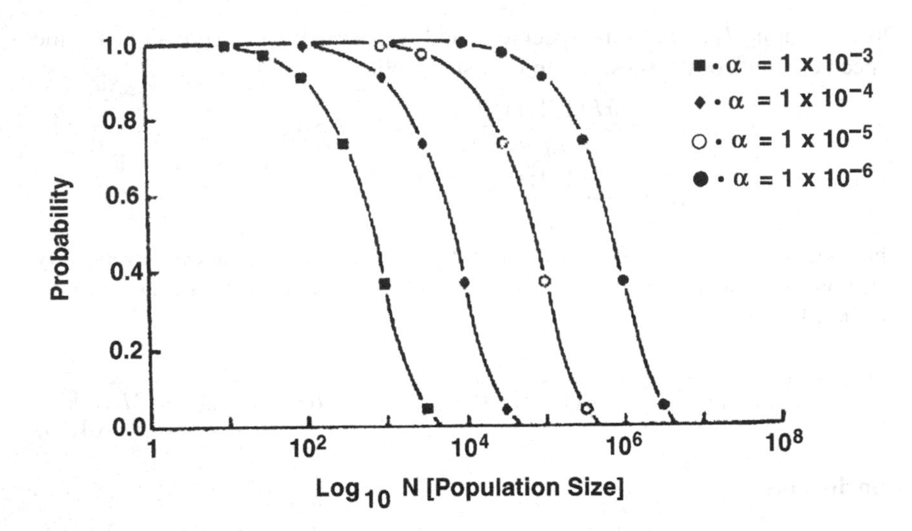

FIGURE 4.3. Probability $P(t)$ of no resistant cells depending on mutation rate α and tumor size $N(t) = \exp(\lambda t)$ in the single-mutation model. Source: Coldman, A.J. and Goldie, J.H. 1987. Modeling resistance to cancer chemotherapeutic agents. Ch. 8, pp. 315–364. In Cancer Modeling (ed.) J.R. Thompson and B.W. Brown. Marcel Dekker, Inc. NY. Figure 1, page 329. Copyright: 1987 Marcel Dekker, Inc.

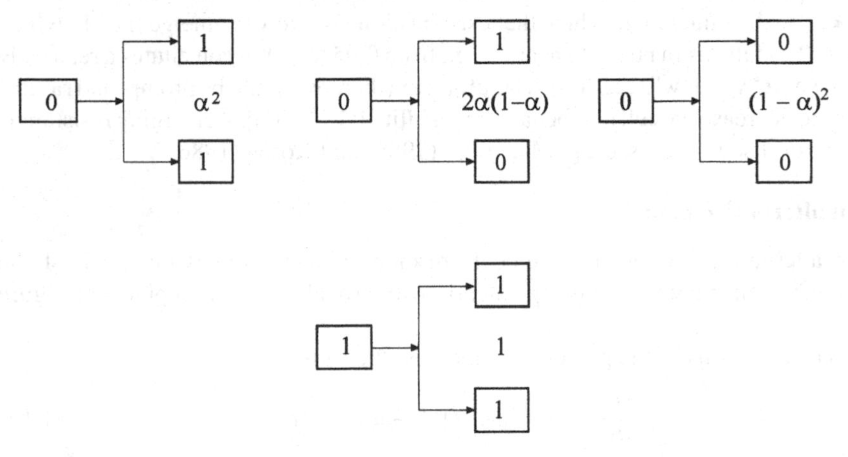

FIGURE 4.4. Schematic representation of the alternative branching process of clonal resistance in the single-mutation case.

the solution of which is

$$P(t) = \frac{1}{[1 - (1 - \alpha)^2]e^{\lambda t} + (1 - \alpha)^2}, \quad t \geq 0. \tag{4.20}$$

If α is small and, consequently, α^2 is a second-order small, then the new $P(t)$ is approximately equal to that in Eq. (4.16), with α replaced by 2α.

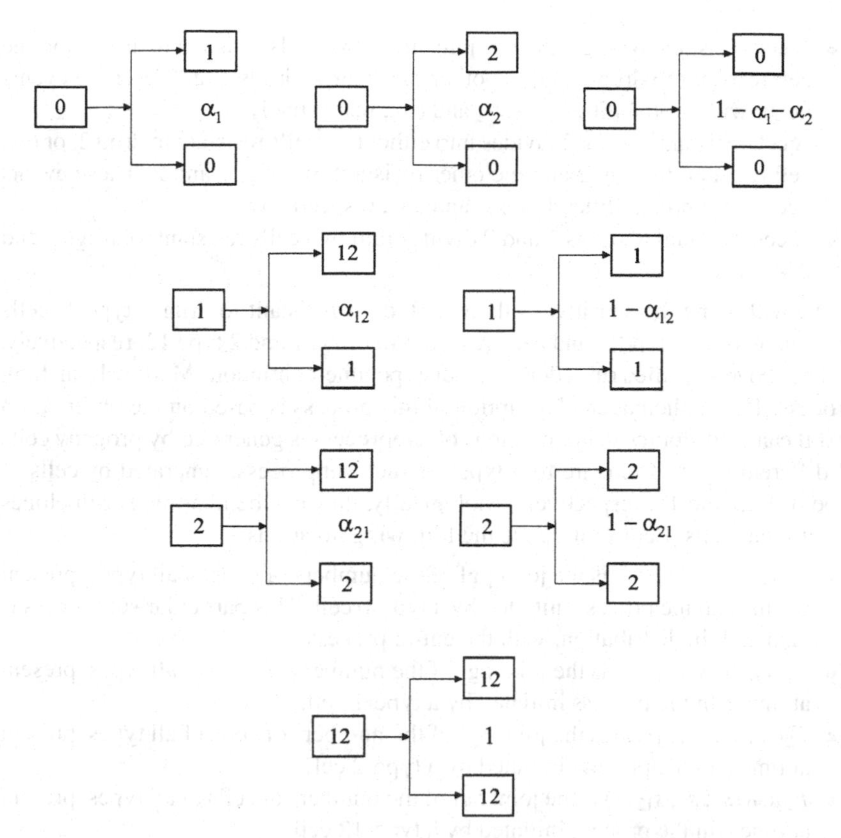

FIGURE 4.5. Schematic representation of the branching process of clonal resistance in the two-mutation case.

4.2.2 *Two-mutation case*

The aim of the two-mutation model is to address the problem of the so-called cross-resistance (i.e., resistance to more than one cancer cell killing agent). Cross-resistance is important for cancer chemotherapy because protocols including more than one agent are frequently used in therapy.

We will specify the hypotheses of our model (Fig. 4.5):

- The population of cells proliferates by binary fission starting from a single cell. The life spans of all the cells are independent exponentially distributed random variables with parameter λ.
- The founder cell of the population is sensitive to chemotherapy.
- A sensitive cell divides into either two sensitive cells, or one sensitive cell and the other cell resistant to drug 1, or one sensitive cell and the other cell resistant to drug 2. These events occur with probabilities $1 - \alpha_1 - \alpha_2$, α_1, and α_2, respectively.

- A cell resistant to drug 1 divides into either two cells resistant to drug 1, or one cell resistant to drug 1 and the other resistant to drugs 1 and 2. These events occur with probabilities $1 - \alpha_{12}$ and α_{12}, respectively.
- A cell resistant to drug 2 divides into either two cells resistant to drug 2, or one cell resistant to drug 2 and the other resistant to drugs 1 and 2. These events occur with probabilities $1 - \alpha_{21}$ and α_{21}, respectively.
- A cell resistant to drugs 1 and 2 divides into two cells resistant to drugs 1 and 2.

We will name the sensitive cells type 0, cells resistant to drug 1 type 1, cells resistant to drug 2 type 2, and cells resistant to drugs 1 and 2 type 12, respectively.

The above-specified rules define a four-type time-continuous Markov branching process. The mathematical description of this process is based on the observation that it can be decomposed into unions of subprocesses generated by progeny cells of different types. There are four types of such subprocess, generated by cells of type 0, 1, 2, and 12, respectively. Biologically, they can be identified with clones of different cells. Let us introduce the following notations:

- $F_0(s_0, s_1, s_2, s_{12}; t)$ is the joint pgf of the numbers of cells of all types, present at time t in the process initiated by a type-0 cell. This particular subprocess is identical, in distribution, with the entire process.
- $F_1(s_0, s_1, s_2, s_{12}; t)$ is the joint pgf of the numbers of cells of all types, present at time t in the process initiated by a type-1 cell.
- $F_2(s_0, s_1, s_2, s_{12}; t)$ is the joint pgf of the numbers of cells of all types, present at time t in the process initiated by a type-2 cell.
- $F_{12}(s_0, s_1, s_2, s_{12}; t)$ is the joint pgf of the numbers of cells of all types, present at time t in the process initiated by a type-12 cell.

We obtain the following system of ordinary differential equations for the probability generating functions F_0, F_1, F_2, and F_{12}:

$$\frac{dF_0}{dt} = -\lambda F_0 + \lambda[(1 - \alpha_1 - \alpha_2)F_0^2 + \alpha_1 F_0 F_1 + \alpha_2 F_0 F_2], \tag{4.21}$$

$$\frac{dF_1}{dt} = -\lambda F_1 + \lambda[(1 - \alpha_{12})F_1^2 + \alpha_{12} F_1 F_{12}], \tag{4.22}$$

$$\frac{dF_2}{dt} = -\lambda F_2 + \lambda[(1 - \alpha_{21})F_2^2 + \alpha_{21} F_2 F_{12}], \tag{4.23}$$

$$\frac{dF_{12}}{dt} = -\lambda F_{12} + \lambda F_{12}^2. \tag{4.24}$$

The initial conditions are $F_i(s; 0) = s_i$, $i = 0, 1, 2, 12$, where $s = (s_0, s_1, s_2, s_{12})$.

It is a little surprising that there exists a semiexplicit solution of this problem. Equation (4.24) can be solved by separation of variables. It yields

$$F_{12}(s; t) = \frac{1}{1 - (1 - s_{12}^{-1})e^{\lambda t}}. \tag{4.25}$$

Substituting expression (4.25) into Eqs. (4.22) and (4.23) and solving the resulting differential equations by separation of variables or application of Theorem 12, yields, respectively,

$$F_1(s;t) = \frac{e^{-\lambda t}[e^{-\lambda t}s_{12} + (1 - s_{12})]^{-\alpha_{12}}}{s_1^{-1} + \{[e^{-\lambda t}s_{12} + (1 - s_{12})]^{1-\alpha_{12}} - 1\}s_{12}^{-1}}, \tag{4.26}$$

$$F_2(s;t) = \frac{e^{-\lambda t}[e^{-\lambda t}s_{12} + (1 - s_{12})]^{-\alpha_{21}}}{s_1^{-1} + \{[e^{-\lambda t}s_{12} + (1 - s_{12})]^{1-\alpha_{12}} - 1\}s_{12}^{-1}}. \tag{4.27}$$

Following the substitution of Eqs. (4.25)–(4.27), Eq. (4.21) assumes the form which is solvable using Theorem 12. Accordingly, we calculate

$$e^{\int_0^t f(u)\,du} = e^{-\lambda t}\{1 + \{[e^{-\lambda t}s_{12} + (1 - s_{12})]^{1-\alpha_{12}} - 1\}s_1 s_{12}^{-1}\}^{\alpha_1/(1-\alpha_{12})}$$
$$\times \{1 + \{[e^{-\lambda t}s_{12} + (1 - s_{12})]^{1-\alpha_{21}} - 1\}s_2 s_{12}^{-1}\}^{\alpha_2/(1-\alpha_{21})}. \tag{4.28}$$

Unfortunately, $\int_0^t e^{\int_0^u f(v)dv}\,du$ cannot be obtained in a closed form. However, we are mainly interested in the probability that no doubly resistant cells emerge before t in the subprocess initiated by a sensitive cell,

$$P_{12}(t) = \mathrm{P}\{N_{12}(t) = 0\} = F_0(1, 1, 1, 0; t). \tag{4.29}$$

In this special case, expression (4.28) is reduced to

$$e^{\int_0^t f(u)\,du} = e^{-\lambda t}[\alpha_{12} + (1 - \alpha_{12})e^{-\lambda t}]^{\alpha_1/(1-\alpha_{12})}$$
$$\times [\alpha_{21} + (1 - \alpha_{21})e^{-\lambda t}]^{\alpha_2/(1-\alpha_{21})}. \tag{4.30}$$

The closed-form solution is still not available, although numerical quadrature is straightforward. However, there exists a special case of interest in which the closed-form solution is available.

- Suppose that all the mutation probabilities are equal (i.e. $\alpha_1 = \alpha_2 = \alpha_{12} = \alpha_{21} = \alpha$).

  In this case,

$$P_{12}(t) = \frac{e^{-\lambda t}[\alpha + (1 - \alpha)e^{-\lambda t}]^{-2\alpha/(1-\alpha)}}{1 - [(1 - 2\alpha)/(1 - 3\alpha)]\{1 - [\alpha + (1 - \alpha)e^{-\lambda t}]^{(1-3\alpha)/(1-\alpha)}\}}. \tag{4.31}$$

Conclusions

Based on the model, the following observations can be made:

- For different mutation rates α, with α small, the plots of $P(t)$ are merely shifted.
- The time interval in which cross-resistance is likely to emerge [i.e., in which $P_{12}(t)$ falls from near 1 to near 0, e.g., from 0.95 to 0.05], constitutes a relatively short "window," similar to that in Figure 4.3.
- It can be proved, similarly to the one-mutation model, that the average number of cells resistant to any of the agents separately increases exponentially. Suppose that we have, at our disposal, agents 1 and 2 and that we can use them according to any time schedule, provided they are not used simultaneously. Because in practice, only periodic chemotherapy protocols are administered, the question is, should the two drugs be alternated frequently or infrequently?

The probability of double resistance emerging from cells resistant to agent 1 strongly depends on the total number of these cells. Therefore, while using agent 1, the number of cross-resistant cells should be kept in check. This is more difficult if agent 1 is used for a long period without a break. The reason is that the cells resistant to agent 1 grow to large numbers, increasing the probability of cross-resistance.

- Summarizing, the two agents should be alternated as frequently as possible. This is the conclusion of Goldie et al. (1982). For a discussion, see Kuczek and Chan (1988).

The original analysis of Goldie et al. (1982), replicated in this section, made use of the simplifying assumption that the two agents were equivalent in their cell killing efficiency (i.e., $\alpha_1 = \alpha_2 = \alpha_{12} = \alpha_{21} = \alpha$). Day (1986a) confirmed the results of Goldie et al. (1982) and extended their analysis by relaxing the symmetry assumption. He analyzed the relative effect of strategies that use agents with different kill efficiencies by using a continuous-time stochastic birth–death multitype branching process model (Day, 1986b) and simulation. The strategies he analyzed included alternating agents, interweaving but not strictly alternating strategies, and one-agent strategies. With each strategy, a wide range of parameters were considered, including treatment scheduling times, cell doubling times, cell mutation rates, drug kill efficiencies, and single resistance or cross-resistance. The simulation results suggest two surprising conclusions: (1) When using two drugs, it is best to use the least effective drug first, or for a longer duration; (2) for some values of tumor kinetics and drug kill parameters, nonalternating treatment can out perform alternation and combination treatment schedules. Practical application of these analyses depends on knowing the appropriate drug kill parameters for each tumor of each patient, although simulation results provide guidelines in the absence of a knowledge of exact parameter values. A different approach to optimization of chemotherapy can be found in Kimmel and Swierniak (1982), Polanski et al. (1993, 1997), Swierniak and Kimmel (1984, 1991), and Swierniak et al. (1996).

4.3 Genealogies of Branching Processes

One of the interesting issues concerning a branching process is that of its genealogy. In broad terms, given a sample of individuals alive at given time t, we trace their past epochs of branching (i.e., the nearest common ancestors of the sample). This exercise is nontrivial because the sample we deal with consists of individuals with positively biased life lengths. This latter effect is due to length-biased sampling analogous to that known from renewal processes. Our treatment is based on the article by O'Connell (1995) (also, see O'Connell, 1993). It concerns the processes close to the critical process, in which the random effects are most pronounced. The theoretical results developed here are illustrated in Section 4.4 by an application concerning the estimation of the age of the common female ancestor of modern humans.

4.3.1 "Near-critical" processes

We consider a family of time-continuous Markov branching processes (age-dependent processes with exponential lifetimes) parameterized by $t \geq 0$. Let $Z_t(u)$ be such a process with mean lifetime 1 and offspring distribution ξ_t with $\mathrm{E}(\xi_t) = 1 + \alpha/t + o(1/t)$ and $\mathrm{Var}(\xi_t) = \sigma^2 + o(1/t) < \infty$, where $\alpha \in R\backslash\{0\}$. We assume that $\alpha \neq 0$ for notational convenience only; the corresponding results for the critical case can be extrapolated by letting $\alpha \to 0$. For this reason, we refer to it as the near-critical case. We will consider the genealogy of this process for fixed α and large t. A general reference on near-critical branching processes is the book of Jagers (1975, pp. 63–70, 199–206). We denote by $\mathrm{P}^x = \mathrm{P}^x_t$ the law of the process Z_t started at x, suppressing the subscript for notational convenience, and write E^x for the corresponding expectations. Set $f_t(s) = \mathrm{E}^1(s^{Z_t(1)})$. It is important to note (see, e.g., Harris, 1963) that the embedded (discrete-time) process $\{Z_t(n), n \in Z_+\}$ is a Galton–Watson process with offspring mean $1+\alpha/t+o(1/t)$, variance $\sigma^2 + o(1/t)$, and pgf $f_t(s)$. For $r > 0$, set $p_{x,r,t} = \mathrm{P}^x[Z_t(rt) > 0]$. We will assume throughout this section that $(Z_t^2(1)|Z_t(0) = 1)$ is uniformly integrable in t.

The first result describes the rate at which $p_{x,r,t} \to 0$ when $t \to \infty$ and an exponential limit law for near-critical Markov branching processes.

Theorem 13.

1. *As* $t \to \infty$, $p_{x,r,t} \sim a_r x/t$, *where*
$$a_r = \frac{2\alpha}{\sigma^2}(1 - e^{-\alpha r})^{-1}.$$
2. *If* $Z_t(0)/t \Rightarrow 0$ *as* $t \to \infty$, *then for* $\lambda > 0$, $x \in Z_+\backslash\{0\}$,
$$\mathrm{E}^x[\exp\left(\frac{-\lambda Z_t(rt)}{t}\right)|Z_t(rt) > 0] \to \frac{b_r}{b_r + \lambda}, \quad t \to \infty;$$
That is,
$$\mathrm{P}\left\{\left[\frac{Z_t(rt)}{t}\right] > z|Z_t(rt) > 0\right\} \to \exp(-b_r z), \quad t \to \infty,$$
where $b_r = e^{-\alpha r} a_r$. *The limit law is exponential with parameter* b_r.

The proof of the theorem is based on a direct diffusion approximation of the branching process. The next result concerns the process reduced to individuals having living descendants.

For each t and $0 \leq u < t$, define the reduced process $N_t(u)$ to be the number of individuals in the process Z_t alive at time u and having descendants alive at time t. Note that for each t, N_t is a time-inhomogeneous Markov branching process. In the statement of the theorem, $D_{Z_+}[0, 1)$ denotes the space of càdlàg (continuous from the right, bounded from the left) paths in Z_+, parameterized by the unit interval; the weak convergence in this case requires only convergence of finite-dimensional distributions. The linear pure birth process with jump rate $b(t)$ is a time-continuous Markov chain $\{N(t), t \geq 0\}$, in which $\mathrm{P}[N(t+\Delta t) = N(t)+1] = b(t)N(t) + o(\Delta t)$, where $o(\Delta t)/\Delta t \to 0$, when $\Delta t \to 0$.

Theorem 14. *As $t \to \infty$, the sequence of processes $\{N_t(rt),\ 0 \le r < 1\}$ converges in distribution in $D_{Z_+}[0,1)$ to a linear pure birth process $\{N(r),\ 0 \le r < 1\}$ with jump rate $b(\alpha, r)N(r)$ at time r, where*

$$b(\alpha, r) = \alpha(1 - e^{-\alpha})^{-1}(1 - r)^{-1},$$

provided $N_t(0) \Rightarrow N(0)$. In particular, as $t \to \infty$,

$$\mathrm{P}^x[N_t(rt) = k \mid N_t(0) = 1] \to q_r(1 - q_r)^{k-1},$$

where $q_r = [\exp(-r) - \exp(-\alpha)]/[1 - \exp(-\alpha)]$.

The result which is of most interest to us describes the degree of relationship of two randomly chosen individuals at time t. Let D_t denote the latest time, counting from the beginning of the process, at which the common ancestor of the two individuals exists. The following theorem provides the asymptotic distribution of this time.

Theorem 15. *For $0 \le r < 1$, $x \in Z_+ \backslash \{0\}$,*

$$\lim_{t\to\infty} \mathrm{P}[D_t > rt \mid N_t(0) = x] = \frac{2q_r^x}{(x-1)!}\{(x-1)!(q_r - 1)^{-x} - F(x-1, 1-q_r)\} \tag{4.32}$$

where $F : Z_+ \times (0,1) \to R$ is defined by

$$F(n, y) = \frac{\partial^n}{\partial y^n}\left\{\frac{\log(1-y)}{y^2}\right\}.$$

Proof. The original proof in O'Connell (1995) is a modification of the corresponding result in Durrett (1978) for the critical case. The current proof is a slight modification of O'Connell (1995), which rectifies some inaccuracies in the original version of expression (4.32). Let $P_{t,u,k}$ denote the probability that two individuals chosen randomly at time t have the same ancestor at time u, given $N_t(u) = k$. Let $X_1(u,t), \ldots, X_k(u,t)$ be independent and identically distributed random variables with the same distribution as $(Z_t(u) | Z_t(u) > 0)$. If we let

$$S_k(u,t) = X_1(u,t) + \cdots + X_k(u,t),$$

then

$$P_{t,u,k} = k\mathrm{E}\{[X_1(u,t)/S_k(u,t)]^2\}.$$

By Theorem 13(1), for each i and $0 \le r < 1$, $X_i(rt,t)$ converges in distribution, as $t \to \infty$, to an exponentially distributed random variable with mean b_r^{-1}, which we denote by $X_i(r)$. So, by bounded convergence, we have

$$P_{t,rt,k} \to k\mathrm{E}\{[X_1(r)/S_k(r)]^2\}.$$

as $t \to \infty$, where

$$S_k(r) = X_1(r) + \cdots + X_k(r).$$

Random variable $X_1(r)/S_k(r)$ can be represented as $Z = X/(X+Y)$, where $X \sim \exp(\psi)$ and $Y \sim$ gamma $(\psi, k-1)$, and X and Y are independent. Independently of constant ψ, this ratio has distribution with density $f_Z(z) = (k-1)(1-z)^{k-2}$, $z \in [0, 1]$, and, consequently, $\mathrm{E}(Z)^2 = 2/[k(k+1)]$. This yields

$$\mathrm{E}\left[\frac{kX_1(r)}{S_k(r)}\right]^2 = \frac{2k}{k+1}.$$

Combining this with Theorem 14, we have, as $t \to \infty$,

$$\begin{aligned}
\mathrm{P}[D_t > rt \mid N_t(0) = x] &= \sum_{k=1}^{\infty} P_{t,rt,k}\mathrm{P}[N_t(rt) > k \mid N_t(0) = x] \\
&\to \sum_{k=1}^{\infty} \frac{2}{k+1}\mathrm{P}[N(r) > k \mid N(0) = x] \\
&\to \sum_{k=1}^{\infty} \frac{2}{k+1}\binom{k-1}{x-1} q_r^x (1-q_r)^{k-x},
\end{aligned}$$

However, by the definition of $F(y, n)$, we have

$$\begin{aligned}
F(y, n) = \frac{\partial^n}{\partial y^n}\left[\frac{\ln(1-y)}{y^2}\right] &= -\sum_{k\geq 1} \frac{\partial^n}{\partial y^n}\left(\frac{y^{k-2}}{k}\right) \\
&= \frac{(-1)^{n+1} n!}{y^{n+1}} - \sum_{k\geq 2} \frac{(k-2)!}{(k-2-n)!(k+1)} y^{k-2-n}
\end{aligned}$$

and, consequently,

$$F(1-q_r, x-1) = (x-1)!(q_r-1)^{-x} - \sum_{k\geq 1} \frac{(k-1)!}{(k-x)!(k+1)}(1-q_r)^{k-x},$$

and the result follows.

Remarks

1. The limiting process in Theorem 14 can be represented as a deterministic time change of a (time-homogeneous) Yule process (in this case, a Markov age-dependent branching process, with progeny number equal to two and lifelength being a random variable distributed exponentially with parameter λ). If $\{Y(t), t \geq 0\}$ is a Yule process with branching rate 1, then the process $\{Y[\ln((1-e^{-\alpha})(e^{-r\alpha} - e^{-\alpha}))], 0 \leq r < 1\}$ has the same law as N.
2. It is instructive to derive explicit expressions for O'Connell's (1995) limit distributions $\Phi_x(r) = \lim_{t\to\infty} \mathrm{P}[D_t/t > r \mid N_t(0) = x]$. We obtain

$$\Phi_1(r) = \frac{2q_r}{(1-q_r)^2}(q_r - 1 - \ln q_r),$$

$$\Phi_2(r) = \frac{2q_r}{(1-q_r)^3}(1 - q_r^2 - 2q_r \ln q_r),$$

where $r \in [0, 1]$. Let us note that $\Phi_1(1) = \Phi_2(1) = 0$, but $\Phi_2(0) = \frac{2}{3}$ and $\Phi_1(0) = 1$. The reason is that in case of the process started by $x = 2$ ancestors, there is a probability equal to $\frac{1}{3}$ that two randomly selected descendants are traced to different ancestors.

3. Similar and related results for general branching processes can be found in Sagitov (1989), Taïb (1987), and Zubkov (1975), for branching diffusion processes in Durrett (1978) and Sawyer (1976), and for superprocesses in Dawson and Perkins (1991) and Etheridge (1992). For an excellent review of the vast literature on genealogical processes in population genetics models, see Tavaré (1984).

4.4 *Application*: Estimation of the Age of the Mitochondrial Eve

4.4.1 Population genetic model

One of the applications of O'Connell's (1995) results is the estimation of the age of the process, which is not observable, based on statistics describing the ages of most recent common ancestors (mrca) of the pairs of extant (contemporary) individuals sampled from the process. The time from mrca of two individuals can be measured using divergence (mismatch) between DNA sequences ascertained in these individuals. O'Connell (1995) presents such an analysis leading to an estimate of the time when the female ancestor of modern humans (mitochondrial Eve, mtEve) lived. We provide an account of his methodology. The estimates which were obtained differ from those obtained using more accepted methods like the coalescence theory (Griffiths and Tavaré 1999). However, the originality of O'Connell's (1995) approach is sufficient to justify this presentation.

Wilson and Cann (1992) and Vigilant et al. (1989, 1991) were the first to hypothesize that a female ancestor of modern humans probably lived in Africa about 200,000 years ago. Hasegawa and Horai (1990) estimate the age to be equal to 280,000 years. Stoneking et al. (1992) published an estimate of 135,000 years. For other more recent examples of estimation of past demographic trends, see, for example, Harpending et al. (1998) and Kimmel et al. (1998).

The data used by O'Connell (1995) are a collection of aligned nucleotide sequences, each approximately 600 base pairs (sites) in length, sampled from the hypervariable segment in the control region of the human mitochondrial genome, of 189 individuals from around the world. There are four nucleotides: adenine, thymine, guanine, and cytosine. A typical sequence might be coded as follows:

$$\text{TTCTTTCCATGGGGAAGCAGA} \cdots \text{CCTAACCAGA}.$$

It is accepted that mitochondrial sequences are maternally inherited and that mitochondrial DNA (mtDNA) in the control region is neutral from the standpoint of natural selection. In other words, the specific makeup of mtDNA does not influence an individual's reproductive fitness.

The following model of mutation is known as the Infinite Sites Model (ISM). A substitution occurs if one of the nucleotides in the sequence is replaced by another, and the new sequence is inherited. According to the molecular clock hypothesis, substitutions occur randomly along lineages at a constant rate, and rates along different lineages are the same. The genetic distance, or divergence, between two such sequences is defined to be the proportion of sites at which the sequences differ. Among humans, this is typically less than 5% in the control region of mtDNA. Vigilant et al. (1989) found the average divergence between the humans in their sample and a sample chimpanzee to be about 15%. The divergence rate is very small, so over the time period we are considering here (the post-Eve period), we can assume that each substitution produces a new type (i.e., reverse substitutions do not occur). Thus, if the most recent common ancestor of two individuals died u million years ago, the number of differences between their mtDNA types will be a random variable with distribution approximated by the Poisson distribution with mean $2u\mu$, where μ is the substitution rate (in units of number of substitutions per million years).

Now, suppose that two individuals are sampled randomly from the current population and δ denotes the rate of divergence (in units of percentage divergence per million years). Note that if l denotes the sequence length, then $\delta = 2\mu/l$. If we have a model for the genealogical structure of the population, then the expected amount of divergence between the mtDNA sequences of the two individuals will be equal to the expected time back to the common ancestor of the two individuals (under our model, in units of millions of years), multiplied by the divergence rate, δ.

Assume that the female population size follows a Markov branching process Z_T with mean number of offspring $1 + \alpha/T$, where $T = T_a/\lambda$; T_a is the time to the most recent common ancestor, λ is the mean effective lifetime (or generation time), and $\alpha \in R$ is the growth parameter.

To obtain an indication of how fast the population might have been growing, suppose that the estimate of 200,000 years was correct. Then, a straightforward moment calculation based on this model with offspring variance $\sigma^2 = 2$, mean (effective) lifetime 25 years, and current (effective) female population size 1 billion yields the rough estimate $\hat{\alpha} = 13.7$ [cf. Eq. (4.33]. The estimate $\hat{\alpha}$ is quite insensitive to apparently large adjustments in these values and remains in the "slightly supercritical" framework for quite a wide range.

We will slightly depart from the original method of estimation in O'Connell (1995). We will use the process with the single ancestor, Eve (i.e., with $x = 1$), whereas O'Connell (1995) used processes generated by the (almost surely) two direct descendants of Eve ($x = 2$). The results are almost identical and our method seems simpler. If we start time at the birth of Eve, then $N_T(0) = 1$. Then, $Z_T(T)$ is the current (effective) female population size. Using the approximation results in Theorems 13–15, we can simultaneously estimate α and T, based on the observations $Z_T(T)$ and the average pairwise divergence in a random sample of n contemporary individuals $\bar{d}_n$. We will assume for the moment that the divergence rate δ is known. Denote by λ the mean effective lifetime of an individual. By

Theorem 13(2),

$$\mathrm{E}[Z_T(T) \mid N_T(0) = 1] \simeq \frac{T_\alpha/\lambda}{b_r} = \frac{\sigma^2 T_a}{2\lambda\alpha}(e^\alpha - 1). \tag{4.33}$$

We also have, by Theorem 15,

$$\begin{aligned} \mathrm{E}[\bar{d}_n \mid N_T(0) = 1] &\simeq \delta\lambda \mathrm{E}[T - D_T \mid N_T(0) = 1] \\ &= \delta T_a \left\{1 - \int_0^1 \mathrm{P}[D_T > rT \mid N_T(0) = 1] dr\right\} \\ &\simeq \delta T_a \gamma_1(\alpha), \end{aligned} \tag{4.34}$$

where

$$\gamma_1(\alpha) = 1 - \int_0^1 \Phi_1(r) dr = 1 - 2\int_0^1 \frac{q_r}{(1-q_r)^2}(q_r - 1 - \ln q_r)\, dr. \tag{4.35}$$

One can simplify Eq. (4.35) to get

$$\gamma_1(\alpha) = 1 - 2\alpha^{-1}\int_0^1 \frac{v}{(1-v)^2(v+\kappa)}(v - 1 - \ln v)\, dv, \tag{4.36}$$

where

$$\kappa = \frac{e^{-\alpha}}{1 - e^{-\alpha}}. \tag{4.37}$$

Note that $\gamma_1(\alpha)$ is positive and increasing in α and that $\gamma_1(\alpha) \uparrow 1$ as $\alpha \to \infty$. For the simplest moment-based estimates, assuming that δ, σ^2, and λ are known, just set

$$Z_T(T) = \frac{\sigma^2 \hat{T}_a}{\lambda\hat{\alpha}}(e^{\hat{\alpha}} - 1), \tag{4.38}$$

$$\hat{T}_a = \frac{\bar{d}_n}{\delta\gamma_1(\hat{\alpha})} \tag{4.39}$$

and solve for $(\hat{\alpha}, \hat{T}_a)$. Although σ^2 is unknown, when α is sufficiently large the actual value (within reason) will not affect the estimates considerably. [This is due to the dominating exponential term in Eq. (4.38).] The same is true for λ.

Remarks concerning performance of the estimators can be found in O'Connell (1995).

4.4.2 *Numerical estimates*

Of the 189 individuals considered by Vigilant et al. (1989), O'Connell (1995) picked a subsample of 19, without being deliberately biased in any way; the sample consists of 6 Asians, 1 Native Australian, 1 Papua New Guinean, 6 Europeans, and 5 Africans. A histogram of the 171 pairwise differences in this sample is shown in Figure 4 of O'Connell (1995). The average divergence was found to be 2.8%.

In June 1992, according to the Population Reference Bureau Estimates, the human population size was approximately 5.412 billion. This gives about 1 billion

TABLE 4.1. Estimates of the Parameters of O'Connell (1995) Model

$\lambda Z_T(T)/\sigma^2$	δ	$\hat{\alpha}$	$\hat{T}_a/10^3$
12.5×10^9	1.8	12.062	1741
	2.7	12.508	1156
	4	12.939	777
5×10^9	1.8	11.047	1761
	2.7	11.497	1168
	4	11.932	785
30×10^9	1.8	13.023	1726
	2.7	13.465	1147
	4	13.893	772

as a rough estimate for the 1992 effective female population size, assuming that about half the population is female and that the current female population represents approximately 2.7 generations. The estimates are quite insensitive to variations in this figure.

Note that the estimates $\hat{\alpha}$ and $\hat{T}_a$ are functions of $\lambda Z_T(T)/\sigma^2$ and δ; these are shown in Table 4.1, for various different values of $\lambda Z_T(T)/\sigma^2$ and δ.

These estimates differ only slightly from the original O'Connell (1995) numbers. If $Z_T(T) = 1$ billion, $\sigma^2 = 2$, and $\lambda = 25$, then $\lambda Z_T(T)/\sigma^2 = 12.5$ billion. Although these choices seem somewhat arbitrary, we can see from Table 4.1 that any kind of realistic deviations from these values will have little or no effect on the estimates. The most important parameter is δ, the rate of divergence.

To derive the estimates for the growth rate, $\hat{\alpha}$, and the age of Eve, $\hat{T}_a$, we simply calculated the expected current population size and the expected average pairwise divergence in a sample of contemporary individuals and assumed that the other parameters were known. We are therefore not fully utilizing the information contained in the sample. It might be helpful to know more about the joint distribution of the pairwise divergences (d_{ij}) or the joint distribution of the respective frequencies of distinct types, in a finite sample. The latter would be analogous to Ewens' sampling formula for the infinite-alleles Wright–Fisher model for neutral evolution (Nagylaki, 1990), which is not applicable to the Eve problem because it is based on the assumption that the population size is constant over time.

4.5 Other Works and Applications

An important application of a branching process involving mutations, similar to the model of Coldman and Goldie (Section 4.2), is the fluctuation test introduced by Luria and Delbrück in 1943 (Luria and Delbrück, 1943). The model and some refinements will be considered in detail in Section 6.1. Here, we will describe the principle and provide a bibliography.

The progeny of a cell may exhibit a new trait that differs from their parent and may pass on the new trait to their own progeny. Let us suppose that the change is due to a single irreversible mutation event. The mutation rate is expressed as the average number of mutations per cell division. Experimentally, a small number of cells is used to seed a series of independent cultures, cells in each culture are allowed to grow, and then the total number of cells in each culture is determined and the number of mutant cells is determined in each culture. The number of cell divisions is estimated from the number of cells in each culture at the beginning and the end of the experiment.

Given parameter values, these models predict the distribution of the number of nonmutant and mutant cells at time t in a population started at time 0 by a single nonmutant cell. In particular, the following observable variables are of interest:

- $N(t)$, the expected total number of non-mutant and mutant cells at time t
- $r(t)$, the expected number of mutant cells at time t.
- $P_0(t)$, the probability of mutant cells being absent from the population at time t

Conversely, given experimental values of $N(t)$, $r(t)$, and $P_0(t)$, it is possible to estimate the parameters of the models – in particular, mutation rates and probabilities.

Models in Section 6.1 illustrate how the estimates obtained differ if alternative assumptions are employed in addition to those originally used by Luria and Delbrück (1943). The literature of the subject includes many more refinements (Kendall and Frost 1988). A review of probability distributions of the number of mutants under differing assumptions can be found in Stewart et al. (1990). Ma et al. (1992), expanded these distributions into a series involving discrete convolution powers. Cell death and differential growth rates were considered in a series of articles by Jones and co-workers (Jones et al. 1994, Jones 1994). Bayesian procedures of estimation of mutation rates were considered by Asteris and Sarkar (1996).

Examples of applications, beyond the original data considered in Luria and Delbrück (1943), will be provided in Section 6.1. They mainly concern mutations to drug resistance in bacteria and cancer cells. One application in a different context is that by Hästbacka et al. (1992), who used a branching process of the Luria and Delbrück type to model the evolution of genetic disease and estimate the location of the disease gene.

An excellent review of various mathematical properties and approximations for the Luria and Delbrück distributions arising from the fluctuation analysis is provided by Angerer (2001). Other, approaches to modeling and estimation of mutation rates are Crump and Hoel (1974) and Tan (1982, 1983).

4.6 Problems

1. *Cells with Exponentially Distributed Lifetimes.* Consider the Markov time-continuous branching process with mean particle lifetime $1/\lambda$. Assume that at its death, each particle produces two specimens of progeny and that each

of them survives, independently, with probability p. Find $h(s)$ and $F(s;t)$. *Consider the critical case separately.*

2. *Problem 1 Continued.* In the critical case, find the limit distribution of $\{\frac{Z(t;\omega)}{t}|Z(t;\omega) > 0\}$, as $t \to \infty$. Compare the result with the corresponding general result for the Galton–Watson process. *Hint:* Consider the Laplace transform
$$\frac{F(e^{-u/t};t) - F(0;t)}{1 - F(0;t)}$$
and use the results of the preceding problem.

3. *Explosions.* Consider the following branching process:
   - A single particle is born at $t = 0$. It lives one unit of time.
   - Each successive generation of particles lives three times shorter than the preceding one.
   - The pgf of progeny number in each generation is $f(s)$ such that $f'(1-) < \infty$.
   - Usual independence hypotheses are verified.

   Find the pgf $F(s,t)$ of $Z(t)$, the number of particles present in the process at time $t \geq 0$. At what time could the process explode? What is the distribution of $Z(t)$ at that time? *Hint.* Consider separately the cases $f'(1-) \leq 1$ and $f'(1-) > 1$.

4. $\{X_n;\ n = 1, 2, \ldots\}$ is a sequence of iid non-negative random variables. Using the weak law of large numbers, demonstrate that
$$\lim_{n\to\infty} \mathrm{P}\{X_1 + X_2 + \cdots + X_n > t\} = 1 \quad \text{for any } t > 0.$$
Hint: Assume first that $\mathrm{E}(X_1) < \infty$. If $\mathrm{E}(X_1) = \infty$, consider truncated rv's $Y_n = \min\{a, X_n\}$, where a is a constant.

5. *Clonal Resistance Revisited.* Consider the following version of the clonal resistance theory.

   a. In the process, there exist two types of particles, labeled 0 (sensitive) and 1 (resistant).
   b. The process is initiated by a single type-0 particle.
   c. The life spans of particles are exponentially distributed independent random variables, with parameter λ.
   d. Each particle, at death, gives birth to exactly two progeny particles:
   - A 0-particle produces either two 0-particles, wp $1 - \alpha$, or two 1-particles, wp α.
   - A 1-particle produces two 1-particles.

   Find the equations for the pgf's $F_0(s_0, s_1;t)$ and $F_1(s_1;t)$. Find and solve the equations for the expected counts of sensitive and resistant cells at time t in the population started at time 0 by a single sensitive cell. Find and solve the equation for $P(t)$, the probability of no resistant cells at time t. Does the change in hypotheses alter the predictions of the theory?

6. *Serial Mutations.* Consider the following branching process:
   a. In the process, there exist three types of particles, labeled 0, 1, and 2.
   b. The process is initiated by a single type-0 particle.
   c. The life spans of particles are exponentially distributed independent random variables, with parameter λ.
   d. Each particle, at death, gives birth to exactly two progeny particles:
   - Each i-particle, $i = 0, 1$, produces either two i-particles, wp $1 - \alpha$, or one i particle and one $i + 1$-particle, wp α.
   - A 2-particle produces two 2-particles.

   The equations for the pgf's $F_0(s_0, s_1, s_2; t)$, $F_1(s_1, s_2; t)$, and $F_2(s_2; t)$ (the joint pgf's of the numbers of the 0-, 1-, and 2-particles, in the process initiated by a single 0-, 1-, and 2-particle, respectively have the following form:

$$\dot{F}_0 = -\lambda F_0 + \lambda[\alpha F_0 F_1 + (1 - \alpha)F_0^2],$$
$$\dot{F}_1 = -\lambda F_1 + \lambda[\alpha F_1 F_2 + (1 - \alpha)F_1^2],$$
$$\dot{F}_2 = -\lambda F_2 + \lambda F_2^2,$$

   with initial conditions $F_0(s_0, s_1, s_2; 0) = s_0$, $F_1(s_1, s_2; 0) = s_1$ and $F_2(s_2; 0) = s_2$. Find and solve the systems of equations for $P_1(t)$ and $P_2(t)$, the probabilities of no 1- and 2-cells at time t, respectively, in the process initiated by a single 0-particle. Draft the plots of $P_1(t)$ and $P_2(t)$. Conclusions?
7. Consider the time-continuous branching process with particle lifetimes distributed exponentially with expectation $1/\lambda$, started by a single ancestor. Assume that at its death, each particle produces two specimens of progeny with probability p and no progeny with probability $1 - p$.
   a. Find $f(s)$.
   b. In the critical case, find $F(s; t)$ and $P(t) = \mathrm{P}\{Z(t, \omega) = 0\}$.
   c. In the critical case, find the limit distribution of

$$\left\{ \frac{Z(t; \omega)}{t} | Z(t; \omega) > 0 \right\},$$

   as $t \to \infty$. Compare the result with the corresponding general result for the Galton–Watson process. *Hint:* Consider the Laplace transform

$$\frac{F(e^{-u/t}; t) - F(0; t)}{1 - F(0; t)}.$$

CHAPTER 5

The Bellman–Harris Process

The Bellman–Harris branching process is more general than the processes considered in the preceding chapters. Lifetimes of particles are non-negative random variables with arbitrary distributions. It is described as follows. A single ancestor particle is born at $t = 0$. It lives for time τ, which is a random variable with cumulative distribution function $G(\tau)$. At the moment of death, the particle produces a random number of progeny according to a probability distribution with pgf $f(s)$. Each of the first-generation progeny behaves, independently of each other and the ancestor, as the ancestor particle did [i.e., it lives for a random time distributed according to $G(\tau)$ and produces a random number of progeny according to $f(s)$]. If we denote the particle count at time t by $Z(t)$, we obtain a stochastic process $\{Z(t),\ t \geq 0\}$. This so-called age-dependent process is generally non-Markovian, but two of its special cases are Markovian: the Galton–Watson process and the age-dependent branching process with exponential lifetimes. The Bellman–Harris process is more difficult to analyze, but it has many properties similar to these two processes.

Frequently, it is assumed that the distribution of lifetimes does not have an atom at $\tau = 0$ [i.e., that $G(0+) = 0$, which is satisfied, among others, when the distribution has a density $g(\tau)$]. This assumption prevents the process from producing infinitely many generations of particles in zero time. The assumption is not always required.

5.1 Integral Equations for the pgf and Basic Properties

We provide a heuristic derivation of the integral equation for the pgf of the number of particles in the Bellman–Harris process $Z(t)$. Because this is one of the most important equations in the theory of branching processes and because it has ramifications in some other branches of applied mathematics (renewal theory, deterministic population dynamics, and others), we also provide a complete derivation in Appendix B.2.

Let us assume that the ancestor's lifetime is equal to τ. Then, for times $t < \tau$, the process consists of a single particle, the ancestor. For times $t \geq \tau$, the number of particles in the process is equal to the sum of the numbers of particles in all subprocesses started by the first-generation progeny of the ancestor; that is,

$$Z(t) = \begin{cases} 1 & t < \tau \\ \sum_{i=1}^{X} Z^{(i)}(t-\tau), & t \geq \tau, \end{cases}$$

where X is the number of the first-generation progeny of the ancestor and $Z^{(i)}(t-\tau)$ are the iid copies of the process, started by these progeny particles at time τ. Denoting the pgf of $Z(t)$ by $F(t, s)$, we obtain in terms of pgf's, conditional on τ:

$$F(s,t) = \begin{cases} s, & t < \tau \\ f[F(s,t-\tau)], & t \geq \tau. \end{cases} \tag{5.1}$$

Removing conditioning (i.e., integrating with respect to the distribution G), we obtain

$$F(s,t) = s[1 - G(t)] + \int_{[0,t]} f[F(s,t-u)]\, dG(u). \tag{5.2}$$

This latter equation is identical to Eq. (B.6) in Appendix B.2.

In general, it is impossible to find explicit solutions of the integral equation (5.2). However, some special cases of interest are described by simpler equations.

Example 1. *Galton–Watson process.* Suppose that $G(t) = \chi(t - T)$, where $\chi(t)$ is the unit step function at 0 (i.e., lifelengths of all particles are identical and equal to T). Equation (5.1) [as well as the integral Eq. (B.6)] now assumes the form

$$F(s,t) = \begin{cases} s, & t \in [0,T) \\ f[F(s,t-T)], & t \in [T,\infty). \end{cases} \tag{5.3}$$

This implies that $F(s,t) = f_n(s)$ if $t \in [nT, (n+1)T)$ and also that $\{Z(nT),\ n = 0, 1, \ldots\}$ is a Galton–Watson process with progeny generating function $f(s)$.

Example 2. *Markov age-dependent branching process.* If we consider the process with life-length distributions $G(u) = 1 - \exp(-\lambda u)$ (i.e., the Markov age-dependent process), then the resulting integral equation can be differentiated side-by-side with respect to t, yielding (after some algebra) the differential equation (4.6).

5.2 Renewal Theory and Asymptotics of the Moments

The theory of the renewal equation plays a major role in investigation of the asymptotic behavior of the Bellman–Harris process. The reason is that the moments of the process are solutions of renewal-type linear integral equations. However, the theory is also important for the nonlinear Eq. (5.2). We will follow Athreya and Ney (1972). Another source is the book by Feller (1971).

5.2.1 *Basics of the renewal theory*

Let us define the renewal function

$$U_m(t) = \sum_{n=0}^{\infty} m^n G^{*n}(t), \quad t \geq 0, \tag{5.4}$$

where G is a distribution on $[0, \infty)$, [i.e., $G(t)$ is non-negative and nondecreasing, $G(\infty) = 1$, and m is a positive constant]. $G^{*n}(t)$ denotes an n-fold convolution of function $G(t)$ by itself [i.e., $G^{*n}(t) = \underbrace{G(t) * \cdots * G(t)}_{n \text{ factors}}$, where $F(t) * G(t) = \int_{[0,t]} F(t-\tau)\, dG(\tau)$ and $F(t)$ and $G(t)$ are bounded nondecreasing functions on $[0, \infty]$.

Lemma 4 (Athreya and Ney 1972). *If* $mG(0+) < 1$, *then* $U_m(t) < \infty$ *for all* $t < \infty$ *and is bounded on finite* t*-intervals.*

Let us consider the renewal equation

$$H(t) = \xi(t) + m \int_0^t H(t-y)\, dG(y), \quad t \geq 0, \tag{5.5}$$

which also can be written as

$$H(t) = \xi(t) + m(H * G)(t),$$

where $\xi(t)$ is a given bounded measurable function on $[0, \infty)$ and $H(t)$ is the unknown function. Let $(\xi * U_m)(t) \equiv \int_0^t \xi(t-y)\, dU_m(y)$ be the convolution of ξ and U_m, which is well defined because U_m is nondecreasing and bounded.

Lemma 5. $H \equiv \xi * U_m$ *is the unique solution of Eq. (5.5), which is bounded on finite intervals.*

The following theorem can be found in Feller's (1971) book. We call a distribution a lattice if its points of increase (or atoms of the corresponding probability measure) occupy isolated points separated by distances being integer multiples of a positive number a. Let us note that if a distribution is nonlattice, then $G(0+) < 1$ is satisfied. The definition of direct Riemann integrability can be found in Feller (1971).

Theorem 16. *Assume* $m = 1$ *and let* $\gamma = \int_0^\infty t\, dG(t) \leq \infty$.

1. *If* $\xi_0 = \lim_{t\to\infty} \xi(t)$ *exists, then the solution of Eq. (5.5) satisfies*

$$\lim_{t\to\infty} \frac{H(t)}{t} = \frac{\xi_0}{\gamma}. \tag{5.6}$$

2. *If ξ is directly Riemann integrable and $G(t)$ is nonlattice, then*

$$\lim_{t\to\infty} H(t) = \frac{1}{\gamma}\int_0^\infty \xi(y)\,dy. \tag{5.7}$$

Definition 4. The Malthusian parameter for m and G is the root, unique provided it exists, of the equation

$$m\int_0^\infty e^{-\alpha y}\,dG(y) = 1. \tag{5.8}$$

We denote it by $\alpha = \alpha(m, G)$.

Let us note that when $m \geq 1$, the Malthusian parameter always exists and is non-negative. If $m < 1$, then α may not exist (if it does, it is negative).

When the Malthusian parameter exists, we can multiply Eq. (5.5) by $e^{-\alpha t}$, and letting

$$H_\alpha(t) = e^{-\alpha t}H(t), \qquad dG_\alpha(t) = me^{-\alpha t}\,dG(t), \qquad \xi_\alpha(t) = e^{-\alpha t}\xi(t),$$

we obtain

$$H_\alpha(t) = \xi_\alpha(t) + \int_0^t H_\alpha(t-y)\,dG_\alpha(y),\ t \geq 0, \tag{5.9}$$

Based on the above, part 2 of Theorem 16 can be used to obtain results of the following type.

Theorem 17. *If the Malthusian parameter $\alpha(m, G)$ exists, if $e^{-\alpha t}\xi(t)$ is directly Riemann integrable, and if G is nonlattice and $mG(0+) < 1$, then the solution H of Eq. (5.5) satisfies*

$$H(t) \sim e^{\alpha t}\left(\int_0^\infty e^{-\alpha y}\xi(y)\,dy\right)\left(m\int_0^\infty ye^{-\alpha y}dG(y)\right)^{-1}. \tag{5.10}$$

5.2.2 *The moments*

In order to derive an equation for the expected number of particles in the process

$$\mathrm{E}[Z(t)] = \mu(t) = \left.\frac{\partial F(s,t)}{\partial s}\right|_{s=1},$$

it is necessary to justify differentiation under the integral in Eq. (5.2). When this is accomplished, the following result is obtained.

Theorem 18. *Suppose $mG(0+) < 1$. $E[Z(t)] \equiv \mu(t)$ is the unique solution of*

$$\mu(t) = m\int_0^t \mu(t-y)dG(y) + 1 - G(t), \tag{5.11}$$

which is bounded on finite t-intervals.

Differentiating the pgf $F(s, t)$ more than once with respect to s, one obtains equations of similar type for higher moments of $Z(t)$. The equation for $[Z(t)] \equiv \mu(t)$ is of the renewal type. Theorem 17, applied to Eq. (5.11) yields the following asymptotic result.

Theorem 19. *Suppose* $mG(0+) < 1$.

1. *If* $m = 1$, *then* $\mu(t) = 1$.
2. *If* $m > 1$ *and* G *is nonlattice, then*

$$\mu(t) \sim ce^{\alpha t}, \quad t \to \infty, \tag{5.12}$$

where α *is the Malthusian parameter for* (m, G) *and*

$$c = \frac{\int_0^\infty e^{-\alpha y}[1 - G(y)]\, dy}{m \int_0^\infty y e^{-\alpha y}\, dG(y)} = \frac{m - 1}{\alpha m^2 \int_0^\infty y e^{-\alpha y}\, dG(y)}. \tag{5.13}$$

3. *If* $m < 1$, *if the Malthusian parameter* $\alpha(m, G)$ *exists, and if* $\int_0^\infty y e^{-\alpha y}\, dG(y) < \infty$, *then relations (5.12) and (5.13) hold, with* $\alpha < 0$.

5.3 Asymptotic Properties of the Process in the Supercritical Case

In the supercritical case, when the Malthusian parameter exists, the asymptotic behavior of the Bellman–Harris process is similar to the behavior of its expected value $\mu(t)$ and to the behavior of the Galton–Watson process. We define the random variable $W(t)$ as

$$W(t) = \frac{Z(t)}{n_1 e^{\alpha t}}, \qquad n_1 = \frac{m - 1}{\alpha m^2 \int_0^\infty u e^{-\alpha u}\, dG(u)}, \tag{5.14}$$

where α is the Malthusian parameter (cf. Definition 4). We see that $\mathrm{E}[W(t)] \longrightarrow 1$, as $t \to \infty$ (cf. Theorem 19).

Theorem 20 (Athreya and Ney 1972). *Suppose that* $m > 1$, $f''(1-) < \infty$, $mG(0+) < 1$ *and* G *is not a lattice distribution. Then,* $W(t)$ *converges with probability* 1 *and in mean squares to a random variable* W, *as* $t \to \infty$, *and*

$$\mathrm{E}(W) = 1, \tag{5.15}$$

$$\mathrm{Var}(W) = \frac{[m + f''(1-)] \int_0^\infty e^{-2\alpha u}\, dG(u) - 1}{1 - m \int_0^\infty e^{-2\alpha u}\, dG(u)}. \tag{5.16}$$

The variance of W *is positive.*

5.4 *Application:* Analysis of the Stathmokinetic Experiment

5.4.1 *Age distributions*

It is frequently necessary to consider the number of particles (objects) not only in the whole process, $Z(t)$, but also in variously defined subsets of the process.

Suppose that for each object in the process, the lifetime τ is the sum of two independent random variables τ_1 and τ_2. This implies $G = G_1 * G_2$, where G_i is the distribution function of τ_i. More specifically, let us assume that the object's life is composed of phase 1 followed by phase 2, with respective durations τ_1 and τ_2. Suppose that we are interested in the number $X(u, t, \omega)$ of objects, at time t, which are in phase 1 and which have time $> u$ remaining to leave phase 1.

An analog of Eq. (5.2) is satisfied by the the pgf $F(s; u, t) = \mathrm{E}[s^{X(u,t)}]$:

$$F(s; u, t) = \int_{0-}^{t+} f[F(s; u, t-\tau)]\, dG(\tau) + s[1 - G_1(t+u)] + [G_1(t+u) - G(t)], \tag{5.17}$$

where $t, u \geq 0$, and $s \in [0, 1]$. Equation (5.17) is of the same type as Eq. (5.2). For a derivation, see Kimmel (1985).

5.4.2 *The stathmokinetic experiment*

The stathmokinetic experiment (Puck and Steffen 1963) was employed by researchers to estimate parameters of cell cycle kinetics [see Darzynkiewicz et al. (1986), for a review]. Basic notions concerning the cell cycle and cell cycle kinetics are explained in Section 2.2. When cell division is blocked by a specific class of external agents, the cells gradually accumulate in mitosis, emptying the postmitotic phase G_1, and with time, also empty the S phase (Fig. 5.1). The pattern of accumulation in mitosis (M) depends on the kinetic parameters of the cell cycle and is used for estimation of these parameters.

The following experimental law is observed in exponentially growing cell populations. Suppose that the cell population grows exponentially as $e^{\lambda t}$. Let us define the collection function $g(t)$ by

$$g(t) = \ln[1 + f_M(t)], \tag{5.18}$$

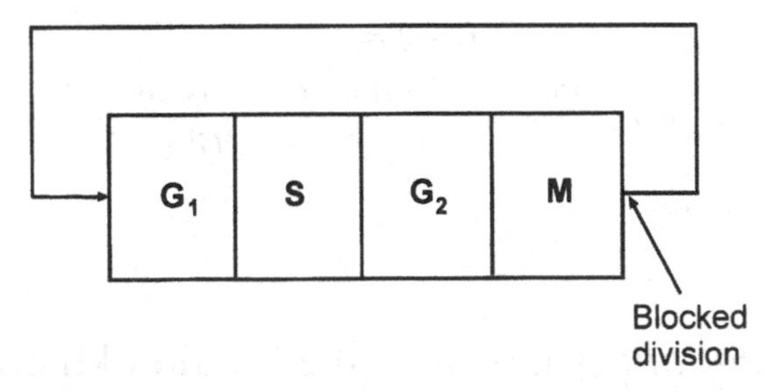

FIGURE 5.1. Generally accepted subdivision of the cell cycle. After division, the daughter cells enter phase G_1, then traverse the phases S, G_2, and M, and then divide. The residence times in all the phases are treated as random. In the stathmokinetic experiment, the divisions are blocked, so that all the cells finally accumulate in M. Source: Kimmel, M. 1985. Nonparametric analysis of stathmokinesis. Mathematical Biosciences 74: 111–123. Figure 1, page 112. Copyright: 1985 Elsevier Science Publishing Co., Inc.

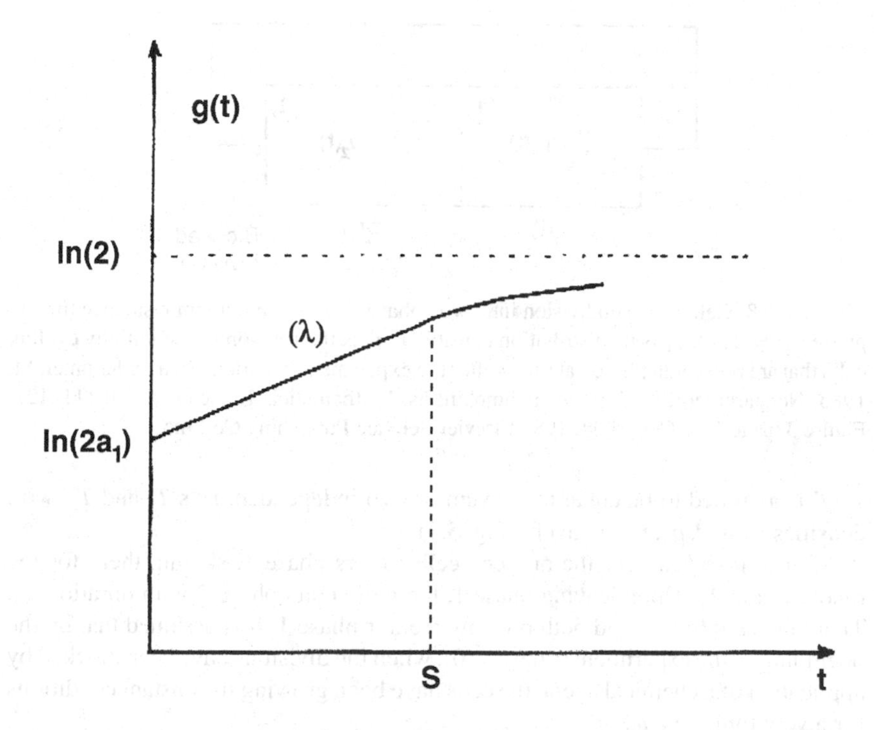

FIGURE 5.2. Typical collection function $g(t)$. S is the minimum residence time in phase 1. For times from the interval $[0, S]$, the collection function is linear with slope λ. Source: Kimmel, M. 1985. Nonparametric analysis of stathmokinesis. Mathematical Biosciences 74: 111–123. Figure 2, page 113. Copyright: 1985 Elsevier Science Publishing Co., Inc.

where $f_M(t)$ is the fraction of the cells in mitosis at time t after starting the experiment. It is frequent that the initial portion of the graph of $g(t)$ is a straight line, the slope of which is equal to λ (Fig. 5.2). Based on this, the growth rate parameter λ, inversely related to the duration of the cell cycle, can be found in an experiment of relatively short duration, in which only the fraction of cells in mitosis is followed. In more sophisticated versions of the stathmokinetic experiment, using the technique called flow cytometry, it is possible to follow fractions of cells in each of the cell cycle phases and, consequently, to estimate more parameters of the cell cycle.

We present a model of the cell cycle based on the Bellman–Harris process. Based on this model, we derive a method of analysis of the stathmokinetic experiment which is independent of the particular functional form of the cell lifetime distribution. The approach follows Kimmel and Traganos (1986) and it is based on previous work by, among others, Jagers and Staudte and their co-workers.

5.4.3 *Model*

It is assumed that proliferating cells follow the rules of a Bellman–Harris process with progeny pgf $f(s) = s^2$. The lifetimes of cells are iid rv's with a generic name

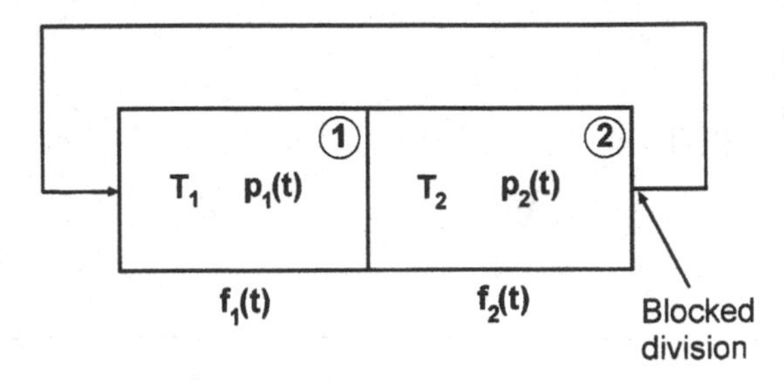

FIGURE 5.3. Cell cycle subdivision into two "phases." T_i is the random residence time in phase i ($i = 1, 2$); p_i is its distribution density; $f_i(t)$ is the fraction of the initially cycling cells that are present in phase i at time t after the experiment is started. Source: Kimmel, M. 1985. Nonparametric analysis of stathmokinesis. Mathematical Biosciences 74: 111–123. Figure 3, page 114. Copyright: 1985 Elsevier Science Publishing Co., Inc.

T. T is assumed to be equal to the sum of two independent rv's T_1 and T_2, with densities p_1 and p_2, respectively (Fig. 5.3).

After mitosis, each of the progeny cells enters phase 1, staying there for the random time T_1. Upon leaving phase 1, the cell enters phase 2 with duration T_2. Then, the cell divides and both progeny reenter phase 1. It is assumed that by the beginning of the experiment (time $t = 0$), when the divisions have been blocked by application of a chemical agent, the cells have been growing in constant conditions for a very long time t_0.

After t_0, when the divisions have been blocked, the total number of cells stays unchanged but the transition from phase 1 to phase 2 continues. Therefore, the number of cells remaining in phase 1 at time t after t_0 is equal to $X(t, t_0)$, as defined in Section 5.4.1. The fraction $f_1(t)$ defined as

$$f_1(t) = \frac{\mathrm{E}[X(t, t_0)]}{\mathrm{E}[Z(t_0)]}, \tag{5.19}$$

where $Z(t_0)$ is the number of cells present at time t_0 (i.e., the number of objects in the Bellman–Harris process), is called the exit curve from phase 1. The accumulation curve in phase 2 is simply $f_2(t) = 1 - f_1(t)$.

Proposition 1. *The exit curve from phase 1 has the asymptotic form*

$$f_1(t) = \lim_{t_0 \to \infty} \frac{\mathrm{E}[X(t, t_0)]}{\mathrm{E}[Z(t_0)]} = 2[1 - P_1(t) - \alpha_1(t)], \tag{5.20}$$

where $P_1(t)$ is the tail distribution function $P[T_1 > t]$ of rv T_1 and

$$\alpha_1(t) = e^{\lambda t} \int_t^\infty p_1(u) e^{-\lambda u}\, du.$$

Here, λ is the Malthusian parameter being the unique real root of the equation

$$2 \int_0^\infty e^{-\lambda y}\, d(P_1 * P_2)(y) = 1.$$

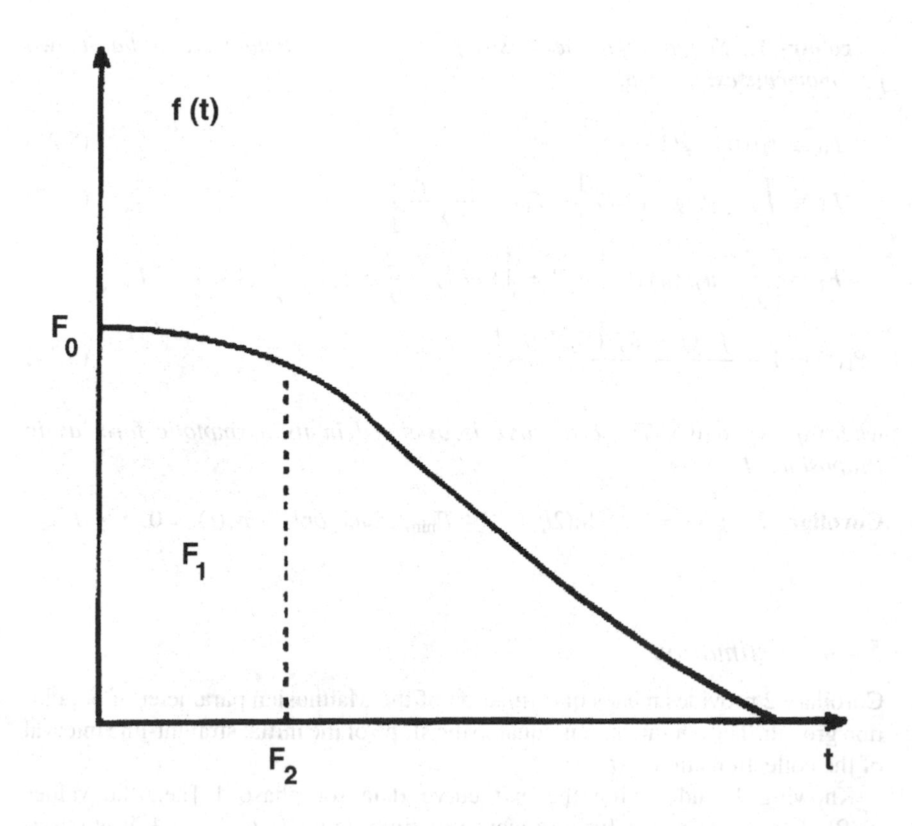

FIGURE 5.4. Typical $f_1(t)$ exit curve. F_0 is the Exponential Steady State (ESS) cell fraction in phase 1; F_1 is the area under $f_1(t)$; F_2 is the coordinate of the mass center of the graph. Source: Kimmel, M. 1985. Nonparametric analysis of stathmokinesis. Mathematical Biosciences 74: 111–123. Figure 4, page 115. Copyright: 1985 Elsevier Science Publishing Co., Inc.

A detailed proof of Proposition 1 can be found in Kimmel (1985). Briefly, to find asymptotics of $f_1(t)$, we have to find the asymptotics of $E[X(t, t_0)]$, as $t_0 \to \infty$. This is done by finding the integral equation for $E[X(t, t_0)]$, which, in turn, is done by employing the pgf equation (5.17). Then, application of the renewal Theorem 17 yields the required asymptotics. An alternative proof could be carried out by using an appropriate random characteristic and Eq. (C.2) of Appendix C.

The following two corollaries describe properties of the exit and collection functions (Fig. 5.4). Proofs can be found in Kimmel (1985). The first corrolary shows that first two moments of the random duration T_1 can be found as solutions of equations involving quantities F_0, F_1, and F_2, which can be computed from the graph of the exit function $f_1(t)$. Also, it shows how to invert the relationship (5.20) in order to compute the tail distribution of T_1, given the exit function. The second corollary demonstrates that the Malthusian parameter is equal to the slope of the linear portion of the accumulation curve $g(t)$.

Corollary 1. *Suppose that the density p_1 exists and is bounded and that its two first moments exist. Then,*

$$F_0 \equiv f_1(0) = 2(1-q), \tag{5.21}$$

$$F_1 \equiv \int_0^\infty f_1(u)\,du = 2\left[\mathrm{E}(T_1) - \frac{1-q_1}{\lambda}\right], \tag{5.22}$$

$$F_2 \equiv \left[\int_0^\infty u f_1(u)\,du\right] F_1^{-1} = \left[\mathrm{E}(T_1^2) - \frac{2}{\lambda}\mathrm{E}(T_1) + \frac{2}{\lambda^2}(1-q_1)\right] F_1^{-1}, \tag{5.23}$$

$$P_1(t) = 1 - \frac{f_1(t) - [df_1(t)/dt]\lambda^{-1}}{2}, \tag{5.24}$$

where $q_1 = \alpha_1(0)$. The exit curve is assumed in its asymptotic form as in Proposition 1.

Corollary 2. $g(t) = \lambda t + \ln(2q_1)$, $t \le T_{\min}$, *if and only if* $p_1(t) = 0$, $t \le T_{\min}$.

5.4.4 *Estimation*

Corollary 2 provides means of estimation of the Malthusian parameter of population growth. The parameter λ is equal to the slope of the initial straight-line interval of the collection curve $g(t)$.

Knowing λ and having the exit curve data for phase 1 [i.e., the values $f_1(0), f_1(t_1), \ldots, f_1(t_n)$, for a sequence of time points $0, t_1, \ldots, t_n$], it is possible to employ Corollary 1 to estimate the duration of phase 1. It can be done in two ways:

1. Calculate from data the "moments" F_0, F_1, and F_2 of the exit curve and solve the three first equations in Corollary 1 for $\mathrm{E}(T_1)$, $\mathrm{E}(T_1^2)$, and q_1.
2. Use the last equation in Corollary 1 to construct a nonparametric estimate of the cumulative distribution $P_1(t)$, based on experimentally recorded values $f_1(0), f_1(t_1), \ldots, f_1(t_n)$, and approximated values $df_1(0)/dt, df_1(t_1)/dt, \ldots, df_1(t_n)/dt$.

A further discussion of applicability of these two methods can be found in Darzynkiewicz et al. (1986) and Kimmel (1985).

The decomposition of the cell cycle into abstract phases 1 and 2 can be carried out in various ways enabling analysis of stathmokinetic data in various biological compartments of the cell cycle (Fig. 5.1). Figure 5.5 depicts the estimation of the first moments of transit times through phases of the cell cycle of the Friend erythroleukemia cells (Kimmel 1985). Figure 6.10a (see Chapter 6) depicts the estimation of the distributions of transit times. For further approaches and applications, see Kimmel et al. (1983, 1986, 1990), and Staiano-Coico et al. (1988).

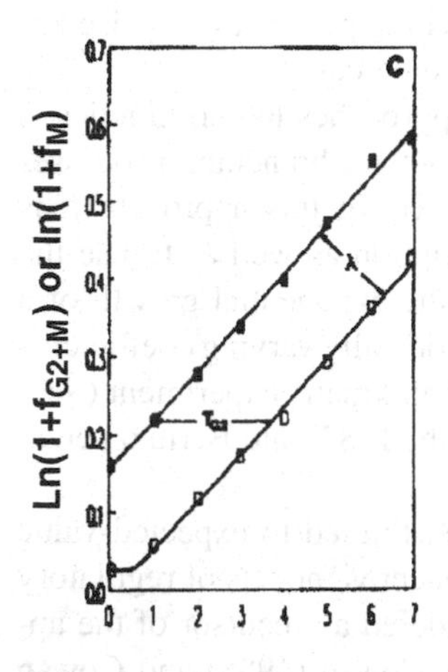

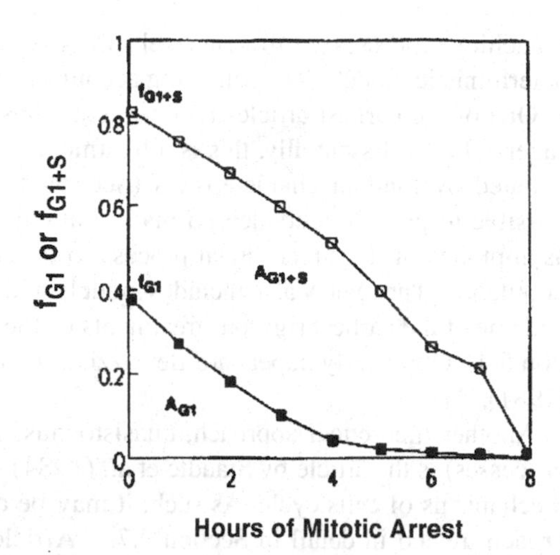

FIGURE 5.5. Analysis of stathmokinetic data for cultured Friend erythroleukemia cells. The first panel depicts the collection curves in phases M and G_2 + M. The (identical) slopes of the straight-line portions of these curves provide the estimate of the Malthusian parameter $\lambda = 0.062$. The second panel depicts the nonparametric estimation of the mean duration of G_1 and S phases. G_1 (closed squares) and G_1 + S (open squares) exit data are depicted in linear scale. The estimate of the mean duration of G_1 is calculated as $E(T_1) = (A_{G_1} + f_{G_1}/\lambda)/2 = (0.82 + 0.38/0.062)/2 = 3.5$ h, where $f_{G_1} = 0.38$ is the fraction of G_1 cells at the beginning of stathmokinesis and $A_{G_1} = 0.824$ is the area under the G_1 exit curve computed from the graph in the second panel based on piecewise linear approximation of the data. The estimate of the mean duration of G_1 + S is calculated as $E(T_1 + T_S) = (A_{G_1+S} + f_{G_1+S}/\lambda)/2 = (3.83 + 0.83/0.062)/2 = 8.6$ h, where $f_{G_1} = 0.83$ is the fraction of G_1 + S cells at the beginning of stathmokinesis and $A_{G_1+S} = 3.83$ is the area under the G_1 + S exit curve. $E(T_S) = E(T_1 + T_S) - E(T_1) = 8.6 - 3.5 = 5.1$ h provides the estimate of the S transit time. Source of first panel: Traganos, F. and Kimmel, M. 1990. The stathmokinetic experiment: A single-parameter and multiparameter flow cytometric analysis. Ch. 25, pp. 249–270. In Methods in Cell Biology, volume 33, Flow Cytometry. (eds) Z. Darzynkiewicz and H.A. Crissman. Academic Press, Inc. San Diego. Figure 3c, page 259. Copyright: 1990 Academic Press, Inc. Source of second panel: Traganos, F. and Kimmel, M. 1990. The stathmokinetic experiment: A single-parameter and multiparameter flow cytometric analysis. Ch. 25, pp. 249–270. In Methods in Cell Biology, volume 33, Flow Cytometry. (eds) Z. Darzynkiewicz and H.A. Crissman. Academic Press, Inc. San Diego. Figure 4, page 261. Copyright: 1990 Academic Press, Inc.

5.5 Other Works and Applications

5.5.1 Cell populations

Cell populations are among natural objects that can be modeled using branching processes and this explains the great number and variety of publications devoted to this subject. A uniform presentation is difficult because different authors employed branching processes at different levels ofor even branching processes disguised as deterministic models. The following account is chronological.

One of the earliest articles reviewing stochastic approaches to cell kinetics is Jagers (1983). Essentially, this is a treatment using general branching processes counted by random characteristics (Section C.1.2). Using this approach, it is possible to provide a condensed mathematical description as well as to use the asymptotics of the supercritical process to describe the exponential growth of a population. The review also includes models with periodically varying coefficients and one of the earliest rigorous treatments of the stathmokinetic experiment (Section 5.4). Other early papers are Bertuzzi and Gandolfi (1983) and Bertuzzi et al. (1981).

Another theoretical approach, quasistochastic (i.e., limited to expected-value processes) is the article by Staudte et al. (1984). It concerns models of regulatory mechanisms of cells cycle. As such, it may be considered a precursor of the approach treated in detail in Section 7.7.1. Articles by Cowan (1985) and Cowan and Morris (1986) belong to the tradition of modeling of the cell cycle using a Bellman–Harris process [also, see Jagers' 1975 book, and Kimmel (1980a, 1980b, 1983)]. To be strict, this should be a multitype process, because cells in different phases of the cell cycle should be considered separately. However, due to the cyclical nature of the problem, it is possible to consider the interdivision time as a convolution of the durations of the successive cell cycle phases. Technically, this is carried out in a way similar to that described in Section 5.4.3.

One of the practical problems for which a mathematical answer is required is how to relate the doubling time t_d of an exponentially growing population (i.e., the time interval needed to increase the mean number of cells by a factor of 2), to the expected life length $\mathrm{E}(T)$ of an individual cell. The exact relationship has the form

$$t_d = \frac{\ln 2}{\alpha},$$

where α is the Malthusian parameter defined as the positive root of the equation

$$m\hat{f}(\alpha) = 1$$

and $\hat{f}(\alpha)$ is the Laplace transform of the density $f(t)$ of the cell life length (Cowan 1985). There is no direct functional relationship between t_d and $\mathrm{E}(T)$. Similarly, there is no direct functional relationship between fractions of cells residing in distinctive phases of the cell cycle and the residence times of cells in these phases. The article by Cowan (1985) provides approximations of the doubling time and of the proportions of cells in different phases in terms of moments of the life length of cells and of the times of residence in cell cycle phases. This leads to a greater insight into the theory and some simple formulas which account for the variability.

In a subsequent article, Cowan and Morris (1986) extend the analysis to the case of cells having different life-length distributions in subsequent generations and becoming quiescent with some probability (possibly different in each generation). This allows modeling of transient effects in differentiating tissues and also of the embryonic phase of the organism's growth [also, see Morris and Cowan (1984) and Morris and Taylor (1985)].

The short book by Knolle (1988) presents the basic ideas of cell proliferation and some mathematical models of population growth. The main application is a cell cycle model with periodic coefficients used for modeling of cancer chemotherapy.

Axelrod et al. (1993) and Gusev and Axelrod (1995) use simulation of branching models to quantify the persistence of cell cycle times of *ras* oncogene-transformed and nontransformed cells over many generations. The experimental system includes primary colonies of cells and secondary colonies grown from cells collected from primary colonies. Persistence of cell cycle times is determined by the heritability of colony sizes (number of cells per colony). The problem of heritability was subsequently studied in more mathematically oriented papers, see Section 6.9.1.

Taïb (1995) studied the functional equation of the form $y'(x) = ay(\lambda x) + b(x)$, which arises in limiting cases of branching models of cell populations. The solution, important for applications, also has an intuitive interpretation as the probability density function of an infinite sum of independent but not identically distributed random variables.

5.5.2 Estimation of cell lifetimes

Estimation of cell lifetimes can be carried out by employing various consequences of the asymptotic balanced exponential growth of the supercritical Bellman–Harris process. The general principle is that the information accumulated in measurable characteristics of the cell population can be disentangled to extract the moments of cell lifetimes, probability of cell death, and so forth. One of the examples is the stathmokinetic experiment of Section 5.4, but other methods also can be used.

Jagers and Norrby (1974) propose a method which involves sampling random cells from an exponentially growing population and following them to division or death. As the sampled cells will usually be of an age greater than 0, the mean t_c of these times is less than the expected cell cycle duration T_c. Indeed,

$$T_c = \frac{1-2p}{2(1-p)}\left(t_c - \frac{T_d}{\ln 2}\right),$$

where T_d is the doubling time of the population and p is the probability of cell death at division. Analogous expressions can be derived for variances. The authors provide statistics to estimate the moments of the cell cycle duration and provide examples of calculations for virally transformed and nontransformed human fetal cell lines. The conclusion is that the transformed cells have longer cell cycle times in spite of shorter population doubling times.

The subject can be treated in more generality. If residence times in different cell cycle phases are random but not independent, then it is necessary to consider the joint probabilities (Macdonald 1978)

$$\psi_i(u_1, \ldots, u_{i-1}, x, y, u_{i+1}, \ldots, u_p)\cdot du_1 \cdots du_{i-1}\cdot dx\cdot dy\cdot du_{i+1} \cdots du_p \quad (5.25)$$

that a cell chosen randomly from the population at time t is in phase i has already spent times $u_1, \ldots, u_{i-1}$ in phases $1, \ldots, i-1$, time x in phase i, and is destined to spend an additional time y in phase i and times $u_{i+1}, \ldots, u_p$ in phases $i+1, \ldots, p$. Although these probabilities are population dependent, the conditional distribution of $y + u_2 + \cdots + u_p$, given $i = 1$ and $x = 0$, is population independent and is the distribution of the life length of a newborn cell. On the contrary, both the backward life length $u_1 + \cdots + u_{i-1} + x$ and its forward counterpart $y + u_{i+1} + \cdots + u_p$, as well as their sum, are population dependent. Distributions $\psi_i(\cdot)$ can be computed under a variety of assumptions. These include the exponential balanced growth, corresponding to the asymptotic behavior of a supercritical process, but also nonstationary cases (varying environment). A review is given by Macdonald (1978). Relationships of this kind allow one to construct correct estimators of quantities more general than those considered by Jagers and Norrby (1974).

An analysis of estimation of mean cell cycle time, based on sample growth trajectories can be found in Hoel and Crump (1974).

Expression (5.25), in the balanced exponential growth version, was used by Cowan and Culpin (1981) to estimate the distribution function of cell residence times in subphases of the cell cycle. The experimental setup was a combination of in vivo fraction-labeled mitoses and arrested division (stathmokinesis) techniques. More specifically, chicken embryo cells were exposed to the 5-bromodeoxyuridine (BUdR), which is incorporated by cells in the S phase of the cell cycle. The amount of BUdR present in the cell is related to the number of times the cell underwent DNA synthesis (i.e., traversed the S phase) during the exposure. Just before the cells were removed from the embryo for measurement, colcemid or colchicine were injected to block further divisions. In this way, more cells accumulate in the the prophase and metaphase (subphases of the M phase) in which the chromosomes can be resolved under the microscope. On the other hand, Macdonald's expression (5.25) makes it possible to calculate the expected numbers of cells that went through a given number of S phases and accumulated in the prophase and metaphase. Using this expression, the model was fitted to observed cell counts, which allowed determining an optimum set of parameters characterizing the durations of cell cycle phases.

A different type of problem is considered in the articles by Axelrod and his co-workers. The main theme is the estimation of parameters of the cell cycle and the modes of dependence between related individuals, based on careful experiments with cell colonies (i.e., clonal cell populations). The first of this series of articles (Axelrod et al. 1986) concerned the distributions of cell life lengths of Friend erythroleukemia cells. In addition to estimates of the α and β curves (tails of the distribution of life lengths and of the distribution of differences between life lengths of sib cells, respectively) the article considers the issue of how dependencies between related cells are altered when cells are treated by cytotoxic agents (in this case, the differentiating agent DMSO). In Friend cells, the α curves become more elongated (i.e., life lengths longer and more dispersed), and the β curves are not altered. This is interpreted as consistent with sib–sib life lengths correlations being increased in treated cells. In later articles, the main subject is the heterogeneity between colonies and its influence on estimated parameters such as correlations between lifetimes of related cells (Kuczek and Axelrod 1986). Kuczek and Axelrod (1987) and Axelrod et al. (1997) introduced a divided colony assay to reduce the influence of heterogeneity on estimations of the influence of cytotoxic drugs on growth of cell colonies [also, see Axelrod and Kuczek (1989)].

5.5.3 *Bifurcating autoregression*

A particularly successful method of estimating parameters of cell proliferation is the bifurcating autoregression developed by Cowan and Staudte (1986). The method applies to branching populations with correlations between relatives defined in an autoregressive manner. In a genealogy of cells, if cell death is excluded, the progeny of a cell m with generation time x_m can be labeled $2m$ and $2m+1$ and their generation times x_{2m} and x_{2m+1}, respectively. The ancestral cell is denoted 1. It is assumed that $x_1 \sim N(\mu, \sigma^2)$ and that, given x_m, the sib times x_{2m} and x_{2m+1} satisfy the relationships

$$x_{2m} - \mu = \theta(x_m - \mu) + e_{2m},$$
$$x_{2m+1} - \mu = \theta(x_m - \mu) + e_{2m+1},$$

where (e_{2m}, e_{2m+1}) are bivariate Normally distributed with common mean zero, common variance λ^2, and correlation coefficient ϕ. From these assumptions, the moments of x_m can be calculated, including the parent–progeny and sib–sib correlations. Consequently, a likelihood function of an observed pedigree can be calculated and numerically maximized to obtain the maximum likelihood estimates of μ, λ^2, and ϕ. The method was modified to accommodate relaxed assumptions and successfully employed to diverse data sets (Staudte 1992, Staudte et al. 1997 and references therein).

5.6 Problems

1. *Geometric Bellman–Harris Process.* Suppose that the particle lifetime distribution is geometric; that is, that $\Pr\{\tau = i\} = (1-p)p^i$, $i \geq 0$ [progeny pgf is a general $h(s)$]. Prove that $\{Z_i,\ i = 0, 1, \ldots\}$, where $Z_i = Z(i)$ is a Galton–Watson process with some progeny pgf $f(s)$ (and consequently that $\{Z_i,\ i = 0, 1, \ldots\}$ is Markov). Find $f(s)$. *Hint:* Write the equation for pgf of Z_i and proceed by induction. Another proof is possible using the lack of memory of the geometric distribution.
2. *The Inverse Problem.* Find the necessary and sufficient condition for a Galton–Watson process with progeny pgf $f(s)$ to be representable as a geometric Bellman–Harris process. *Hint:* Check if $h(s)$ corresponding to a given $f(s)$ is a pgf.

3. *Age Distributions.* Find the integral equation for the pgf $F(s; y, t)$ of $Z(y, t)$ (number of particles at time t, with ages $\leq y$). Use the property

$$Z(y,t) = \sum_{k=1}^{X} Z^{(k)}(y, t-\tau) \quad \text{if } \tau \leq t$$

and reasoning as in the heuristic derivation of Eq. (5.2).

4. *Expected Age Distributions.* Prove that if the Malthusian parameter exists, then, as $t \to \infty$, the normed expected age distribution $A(y, t) = \mathrm{E}[Z(y, t)]/\mathrm{E}[Z(t)]$ tends to the limit

$$A(y) = \frac{\int_0^y e^{-\alpha z}[1 - G(z)]\, dz}{\int_0^\infty e^{-\alpha z}[1 - G(z)]\, dz}.$$

Hint: Find the integral equation for $\mu(y, t) \equiv \mathrm{E}[Z(y, t)]$ and use the asymptotics of Theorem 17.

5. *The $A \to B$ Transition Model of the Cell Cycle.* Suppose that in a proliferating cell population, a newborn cell, with probability p, stays dormant until it is prompted into further growth by a random "hit," which occurs (independently for each cell) with probability $\beta\tau + o(\tau)$ in any short time interval of duration τ. After this "hit," the cell requires a fixed time T to grow and divide. Cells which do not require the "hit" start growing at the moment of birth. No cell death occurs. Find the limit age distribution $A(y)$. If, for a cell population growing long enough, the empirical age distribution can be found, can it help in establishing the value of p (which is a biologically important parameter)?

6. *Bellman–Harris Process, the Lattice Case.* Consider the age-dependent branching process $\{Z_n, n = 0, 1, \ldots\}$ with progeny pgf $h(s)$ and the lifetime distribution $\{g_i, i = 1, ..., k\}$. Prove that the pgf $f_n(s)$ of Z_n is equal to $f_n^1(s, \ldots, s)$, where $f_n^1(\mathbf{s})$ is the pgf of the k-type Galton–Watson process $\mathbf{Z}_n$ (initiated by a single type-1 particle), with the following progeny pgf's:

$$f^i(\mathbf{s}) = (1 - \gamma_i)h(s_1) + \gamma_i s_{i+1}, \quad i = 1, \ldots, k-1,$$
$$f^k(\mathbf{s}) = (1 - \gamma_k)h(s_1),$$

where $\gamma_i = \bar{G}_{i+1}/\bar{G}_i$. (In other words, the total number of particles of all types in this k-type Galton–Watson process is equal to the number of particles in the lattice Bellman–Harris process. What is the interpretation of particle type here?)

7. *Perron-Frobenius Root.* Assume $h'(1-) > 1$. Find the determinant equation for the maximum real eigenvalue ρ of matrix $\mathbf{M}$. Proceed by induction with respect to k. Check that the process is supercritical (i.e., that $\rho > 1$). Find the left eigenvector ν corresponding to ρ.

8. Show that the age distribution of particles in the process Z_n (i.e., the vector $(Z_{n1}, \ldots, Z_{nk})$, where Z_{ni} is the number of particles with age i at time n), has pgf $f_n^1(\mathbf{s})$. Based on this and the limit law for the multitype supercritical positive regular Galton–Watson process, state the limit law for the age distribution of the lattice Bellman–Harris process.

CHAPTER 6

Multitype Processes

In the present chapter, we present models involving branching processes with many types of particle. Multitype models were sporadically employed in previous chapters. Here, we offer a systematic treatment of asymptotic properties of the multitype Galton–Watson process in the supercritical case. However, we start with a motivating application, involving several multitype approaches to the fluctuation experiment analysis, which is one of the oldest but still useful tests of mutagenesis. Other applications follow.

6.1 *Application*: Two-Stage Mutations and Fluctuation Analysis

The progeny of a cell may exhibit a new trait that differs from their parent and may pass on the new trait to their own progeny. Such a change is usually considered to be due to a single irreversible mutation event. However, a possibility exists that the observed change may be due to an event that has a finite probability of being reversible or may be the result of more than one mutational event.

The rate at which mutations occur in populations of cells has been estimated using the fluctuation test introduced by Luria and Delbrück in 1943. The mutation rate is defined as the average number of mutations per cell division. Experimentally, a small number of cells is used to seed a series of independent cultures, cells in each culture are allowed to grow, and then the total number of cells in each culture is determined and the number of mutant cells is determined in each culture. The number of cell divisions is estimated from the total number of cells in each culture at the beginning and the end of the experiment. The mutation rate can be calculated in two ways (viz. from the total number of mutant cells or from the proportion of cultures with no mutant cells).

The two methods of calculating the mutation rate in the Luria and Delbrück fluctuation test do not always agree. This has motivated the investigation of models for determining mutation rates based on the possibility that some changes in inherited traits are due to more than one mutation event and that some events may be reversible.

We present a series of models of cell growth and mutation. The purpose is to model the fluctuation experiment as applied to the analysis of data on the drug resistance of cells. The material is based on the article by Kimmel and Axelrod (1994). The classical fluctuation analysis is based on a model of cell proliferation and single-stage irreversible mutation introduced in Luria and Delbrück (1943). We summarize the hypotheses and predictions of that model and of four other models employing different hypotheses. These models are modifications of the Luria–Delbrück model, including random cell interdivision time, cell death, and two-stage mutations with the first stage being reversible.

Given parameter values, these models predict the distribution of the number of nonmutant and mutant cells at time t in a population started at time 0 by a single nonmutant cell. In particular, the following observable variables are of interest:

- $N(t)$, the expected total number of nonmutant and mutant cells at time t
- $r(t)$, the expected number of mutant cells at time t
- $P_0(t)$, the probability of mutant cells being absent from the population at time t

Conversely, given experimental values of $N(t)$, $r(t)$, and $P_0(t)$, it is possible to estimate the parameters of the models – in particular, mutation rates and probabilities.

6.1.1 Luria–Delbrück model

The hypotheses are as follows (see Fig. 6.1a and Table 6.1):

1. Two types of cells exist in the population: type-0 nonmutant cells and type-1 mutant cells.
2. All cells in the population have interdivision times equal to $\ln 2$.
3. Each cell, at the moment of division, gives birth to two daughter cells. The type of each of these daughters is the same as that of the mother cell.
4. During its lifetime, independently of any other events, a type-0 cell undergoes an irreversible transformation into a type-1 cell with probability $a\Delta t + o(\Delta t)$ in any brief lifetime interval $(t, t+\Delta t)$. The constant a is called the transition or mutation rate. This implies that if the time from birth to mutation is denoted by T, then $\mathrm{P}[T > t] = \max[\exp(-at), \exp(-a \ln 2)]$ [i.e., the mutation may not occur at all wp $\exp(-a \ln 2) = 2^{-a}$].

The analysis of the model carried out originally in Luria and Delbrück (1943) and reworked in Lea and Coulson (1949) is based on the assumption that the population as a whole is large enough to be treated deterministically, whereas

TABLE 6.1. Summary of Hypotheses of the Models Considered

Model	Interdiv. Time	Probability of Cell Death	Number of Stages	Probability of Mutation
Luria–Delbrück	ln(2)	0	1	$a\Delta t$ in $(t, t+\Delta t)$ (irreversible)
Markov branching	1 (expected)	0	1	$a\Delta t$ in $(t, t+\Delta t)$ (irreversible)
Galton–Watson	1	0	1	α per daughter cell (irreversible)
Galton–Watson with cell death	1	δ	1	α per daughter cell (irreversible)
Galton–Watson two-stage	1	0	2	$0 \to 1: \alpha_{01}$ $1 \to 0: \alpha_{10}$ (reversible) $1 \to 2: \alpha_{12}$ (irreversible)

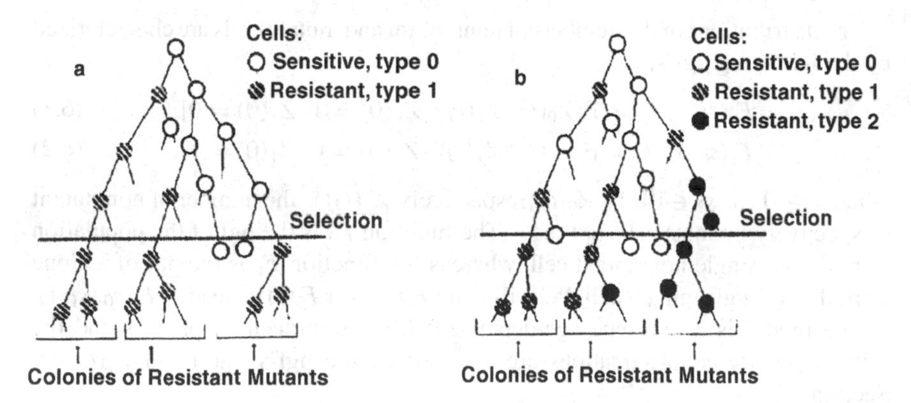

FIGURE 6.1. Schematics of transitions admitted in (a) the one-stage models and (b) the two-stage model. Source: Kimmel, M. and D. Axelrod. 1994. Fluctuation test for two-stage mutations: application to gene amplification. Mutation Research 306: 45–60. Figure 1, page 48. Copyright: 1994 Elsevier Science B.V.

the mutation events are rare and, therefore, the mutants have to be counted in a probabilistic manner. Solutions, which were derived in Lea and Coulson (1949), are listed in the first row of Table 6.2. We do not provide derivations, referring the reader to Kimmel and Axelrod (1994).

TABLE 6.2. Expressions for the Expected Total Count of Cells $N(t)$, for the Expected Count of Mutant Cells $r(t)$ and for the Probabilities of No Mutant Cells $P_0(t)$ in the Models Considered.

Model	$N(t)$	$r(t)$	$P_0(t)$
Luria–Delbrück	e^t	ate^t	$\exp(-ae^t)$
Markov branching process	e^t	$e^t(1-e^{-at})$	$\frac{a+1}{ae^{(a+1)t}+1}$
Galton–Watson process	2^t	$2^t[1-(1-\alpha)^t]$	$(1-\alpha)^{2^{(t+1)}-2}$
Galton–Watson with cell death	$[2(1-\delta)]^t$	$[2(1-\delta)]^t\left[1-\left(\frac{1-\alpha-\delta}{1-\delta}\right)^t\right]$	Eqs. (6.23)–(6.24)
Galton–Watson two-stage	2^t	$2^t-\frac{\rho_1^t}{A_1}-\frac{\rho_2^t}{A_2}$	Eqs. (6.25)–(6.27) and (6.33)

6.1.2 *The Markov branching process model*

In this model, the interdivision time is not constant but is random with exponential distribution. Hypothesis 2 is therefore replaced by the following (cf. Table 6.1 and Fig. 6.1a).

2. All cells in the population have exponentially distributed interdivision times with mean (expected) value equal to 1.

The distributions of the numbers of nonmutant and mutant cells are characterized by the following pgf's:

$$F_0(s_0, s_1; t) = \mathrm{E}[Z_0(t)^{s_0} Z_1(t)^{s_1} \,|\, Z_0(0) = 1,\, Z_1(0) = 0], \tag{6.1}$$

$$F_1(s_0, s_1; t) = \mathrm{E}[Z_0(t)^{s_0} Z_1(t)^{s_1} \,|\, Z_0(0) = 0,\, Z_1(0) = 1], \tag{6.2}$$

where $t \geq 0$, $s_1, s_2 \in [0, 1]$. $Z_0(t)$ [respectively $Z_1(t)$] is the number of nonmutant (respectively mutant) cells at time t. The function F_0 is the pgf of the population started by a single nonmutant cell, whereas the function F_1 is the pgf of a clone started by a single mutant cell. We will write $F_i(s; t)$ or $F_i(t)$ instead of $F_i(s_0, s_1; t)$.

The model is a two-type age-dependent Markov branching process and the following differential equations are satisfied by the pgf's $F_0(t)$ and $F_1(t)$ (cf. Section 4.2.1):

$$\frac{d}{dt}F_0(t) = -(a+1)F_0(t) + (a+1)\left[\frac{1}{a+1}F_0(t)^2 + \frac{a}{a+1}F_1(t)\right], \tag{6.3}$$

$$\frac{d}{dt}F_1(t) = -F_1(t) + F_1^2(t), \quad t \geq 0. \tag{6.4}$$

The initial conditions are $F_i(t) = s_i$, $i = 0, 1$. The form of Eqs. (6.3) and (6.4) can be understood by comparison with Eqs. (4.10) and (4.11). Under the new Hypothesis 2, after a time, which is distributed exponentially with parameter $a+1$, either two type-0 cells are produced [wp $1/(a+1)$] or a single type-1 cell [wp $a/(a+1)$]. This latter cell is a "type-1 continuation" of the type-0 cell.

We are interested in evaluating $N(t)$, $r(t)$, and $P_0(t)$. They can be expressed as

$$N(t) = \mathrm{E}[Z_0(t) + Z_1(t)|Z_0(0) = 1, Z_1(0) = 0]$$
$$= \left(\frac{\partial}{\partial s_0} + \frac{\partial}{\partial s_1}\right) F_0(s;t)\bigg|_{s_0=s_1=1}, \tag{6.5}$$

$$r(t) = \mathrm{E}[Z_1(t)|Z_0(0) = 1, Z_1(0) = 0] = \frac{\partial}{\partial s_1} F_0(s;t)\bigg|_{s_0=s_1=1}, \tag{6.6}$$

$$P_0(t) = F_0(1, 0; t). \tag{6.7}$$

Solving the resulting equations yields the results displayed in Table 6.2.

6.1.3 *The Galton–Watson process model*

In this model, cells mutate immediately following division. Hypotheses 3 and 4 are therefore replaced by the following (cf. Table 6.1 and Fig. 6.1a):

3. Each cell, at the moment of division, gives birth to two daughter cells. The type of each of the daughters may or may not be the same as that of the mother cell.
4. Following division, a type-0 daughter cell undergoes irreversible transformation into a type-1 cell with probability α. The constant α is now called the transition or mutation probability.

The distributions of nonmutant and mutant cells can be characterized by the pgf's $F_0(s_0, s_1; t)$ and $F_1(s_0, s_1; t)$ as defined in Eqs. (6.1) and (6.2), except that the time variable t now assumes only non-negative integer values, equal to the multiples of the interdivision time.

The pgf's $F_0(t)$ and $F_1(t)$ satisfy a system of recurrence equations, stemming from the following vector generalization of the backward iteration (3.2):

$$F(s, t) = h[F(s, t - 1)],$$

where

$$F = (F_0, F_1), \qquad h = (h_0, h_1),$$
$$h_0(s_0, s_1) = [(1 - \alpha)s_0 + \alpha s_1]^2, \qquad h_1(s_0, s_1) = s_1^2.$$

Substituting h_0 and h_1 as given above, we obtain

$$F_0(s;t) = [(1 - \alpha)F_0(s;t - 1) + \alpha F_1(s;t - 1)]^2, \tag{6.8}$$
$$F_1(s;t) = [F_1(s;t - 1)]^2, \tag{6.9}$$

where $s = (s_0, s_1)$, $t = 1, 2, \ldots,$ with initial conditions $F_i(s;0) = s_i$, $i = 0, 1$. Recurrences (6.8) and (6.9) cannot be solved explicitly, but differentiation side-by-side of (6.8) and (6.9) with respect to s_0 and setting $s_0 = s_1 = 1$ yields

$$\mathrm{E}[Z_0(t)|Z_i(0) = \delta_{0i}] = 2\{(1 - \alpha)\mathrm{E}[Z_0(t - 1)|Z_i(0) = \delta_{0i}]$$
$$+ \alpha \mathrm{E}[Z_0(t - 1)|Z_i(0) = \delta_{1i}]\}, \tag{6.10}$$
$$\mathrm{E}[Z_0(t)|Z_i(0) = \delta_{1i}] = 2\mathrm{E}[Z_0(t - 1)|Z_i(0) = \delta_{1i}], \; t = 1, 2, \ldots, \tag{6.11}$$

with initial conditions $E[Z_0(0)|Z_i(0) = \delta_{0i}] = 1$, and $E[Z_0(0)|Z_i(0) = \delta_{1i}] = 0$. This yields $E[Z_0(t)|Z_i(0) = \delta_{1i}] = 0, \ \ t = 0, 1, 2, \ldots$, and

$$E[Z_0(t)|Z_i(0) = \delta_{0i}] = [2(1-\alpha)]^t, \ \ t = 0, 1, 2, \ldots, \tag{6.12}$$

as expected. Because there is no cell death assumed,

$$N(t) = E[Z_0(t) + Z_1(t)|Z_i(0) = \delta_{0i}] = 2^t, \ \ t = 0, 1, 2, \ldots. \tag{6.13}$$

Equations (6.12) and (6.13) yield

$$r(t) = E[Z_1(t)|Z_i(0) = \delta_{0i}] = 2^t - [2(1-\alpha)]^t, \ \ t = 0, 1, 2, \ldots, \tag{6.14}$$

as displayed in Table 6.2.

To obtain $P_0(t)$, we use the definition (6.7) and also denote $P_1(t) = F_1(1, 0; t)$. Substitution of $s_0 = 1$, and $s_1 = 0$ in Eqs. (6.8) and (6.9) yields

$$P_0(t) = [(1-\alpha)P_0(t-1) + \alpha P_1(t-1)]^2, \tag{6.15}$$

$$P_1(t) = [P_1(t-1)]^2, \ \ t = 1, 2, \ldots, \tag{6.16}$$

with initial conditions $P_0(0) = 1$ and $P_1(0) = 0$. Therefore,

$$P_0(t) = (1-\alpha)^{2^{t+1}-2}, \ \ t = 0, 1, 2, \ldots, \tag{6.17}$$

as displayed in Table 6.2

6.1.4 The Galton–Watson process model with cell death

In this model, each of the daughter cells (mutant or nonmutant) may also die with some probability. Hypothesis 4 is therefore replaced by the following (cf. Table 6.1).

4. Following division, a type-0 daughter cell either undergoes irreversible transformation into a type-1 cell with probability α, or dies with probability δ, or stays type-0 with probability $1 - \alpha - \delta$. The type-1 daughter cell may either die with probability δ or stay alive with probability $1 - \delta$.

The presence of cell death leads to the following modification of Eqs. (6.8) and (6.9):

$$F_0(s;t) = [(1-\alpha-\delta)F_0(s;t-1) + \alpha F_1(s;t-1) + \delta]^2, \tag{6.18}$$

$$F_1(s;t) = [(1-\delta)F_1(s;t-1) + \delta]^2, \ \ t = 1, 2, \ldots, \tag{6.19}$$

with initial conditions $F_i(s; 0) = s_i$, $i = 0, 1$. We obtain

$$E[Z_0(t)|Z_i(0) = \delta_{0i}] = [2(1-\alpha-\delta)]^t, \ \ t = 0, 1, 2, \ldots, \tag{6.20}$$

and

$$N(t) = E[Z_0(t) + Z_1(t)|Z_i(0) = \delta_{0i}] = [2(1-\delta)]^t, \ \ t = 0, 1, 2, \ldots, \tag{6.21}$$

which yields

$$r(t) = \mathrm{E}[Z_1(t)|Z_i(0) = \delta_{0i}] = [2(1-\delta)]^t \left[1 - \left(\frac{1-\alpha-\delta}{1-\delta}\right)^t\right], \quad t = 0, 1, 2, \ldots, \tag{6.22}$$

as displayed in Table 6.2. Substitution of $s_0 = 1$ and $s_1 = 0$ in Eqs. (6.18) and (6.19) yields

$$P_0(t) = [(1-\alpha-\delta)P_0(t-1) + \alpha P_1(t-1) + \delta]^2, \tag{6.23}$$

$$P_1(t) = [(1-\delta)P_1(t-1) + \delta]^2, \quad t = 1, 2, \ldots, \tag{6.24}$$

with initial conditions $P_0(0) = 1$, and $P_1(0) = 0$, where $P_0(t)$ is the probability of no mutant cells at time t in the population derived from a nonmutant cell, whereas $P_1(t)$ is the extinction probability (by time t) of a clone started by a mutant. This recurrence has to be solved numerically.

6.1.5 *Two-stage Galton–Watson process model*

In this model, two stages of mutant cells are present: type 1 and type 2. Mutation from type 0 to type 1 is reversible, whereas mutation from type 1 to type 2 is irreversible. Hypothesis 4 is therefore replaced by the following (cf. Table 6.1 and Fig. 6.1b).

4. Following division:

- A type-0 daughter cell undergoes transformation into a type-1 cell, with probability α_{01}.
- A type-1 daughter cell undergoes a reverse transformation into a type-0 cell, with probability α_{10}.
- A type-1 daughter cell undergoes irreversible transformation into a type-2 cell, with probability α_{12}.

The two-stage mutation model is a three-type Galton–Watson process. Its distributions are described by pgf's $F_i(s_0, s_1, s_2; t)$, $i = 0, 1, 2$, where F_0 is the joint pgf of the numbers of cells of types 0, 1, and 2 in the population started by a single nonmutant cell, F_1 is the joint pgf in the population started by a single stage-1 mutant cell, and F_2 is the pgf of the stage-2 mutant clone started by a single stage-2 irreversible mutant. The hypotheses of the model lead to the following recurrent equations for the pgf's:

$$F_0(s;t) = [(1-\alpha_{01})F_0(s;t-1) + \alpha_{01}F_1(s;t-1)]^2, \tag{6.25}$$

$$F_1(s;t) = [\alpha_{10}F_0(s;t-1) + (1-\alpha_{10}-\alpha_{12})F_1(s;t-1) + \alpha_{12}F_2(s;t-1)]^2, \tag{6.26}$$

$$F_2(s;t) = [F_2(s;t-1)]^2, \quad t = 1, 2, \ldots, \tag{6.27}$$

with initial conditions $F_i(s;0) = s_i$, $i = 0, 1, 2$. Let us denote $M(t) = (M_{ij}(t))_{i,j=0,1,2}$, the matrix of expected cell counts

$$M_{ij} = \mathrm{E}[Z_j(t)|Z_k(0) = \delta_{ik}, k = 0, 1, 2] = \frac{\partial F_i(1;t)}{\partial s_j}. \tag{6.28}$$

Differentiating system (6.25)–(6.27), we obtain

$$M(t) = \mu M(t-1), \quad t = 1, 2, \ldots, \tag{6.29}$$

where μ is the expected progeny matrix

$$\mu = 2 \begin{pmatrix} 1-\alpha_{01} & \alpha_{01} & 0 \\ \alpha_{10} & 1-\alpha_{10}-\alpha_{12} & \alpha_{12} \\ 0 & 0 & 1 \end{pmatrix}. \tag{6.30}$$

The initial condition is $M(0) = I$ (the identity matrix). We obtain

$$M(t) = \mu^t, \quad t = 0, 1, 2, \ldots. \tag{6.31}$$

Involved but standard calculations consisting of finding the eigenvalues and eigenvectors of matrix μ lead to the following explicit expression for $r(t)$:

$$r(t) = M_{01}(t) + M_{02}(t) = 2^t - \frac{\rho_1^t}{A_1} - \frac{\rho_2^t}{A_2}, \quad t = 0, 1, 2, \ldots, \tag{6.32}$$

where

$$\rho_i = (2 - \alpha_{01} - \alpha_{10} - \alpha_{12}) + (-1)^i \sqrt{(\alpha_{10} + \alpha_{12} - \alpha_{01})^2 + 4\alpha_{01}\alpha_{10}}$$

and

$$A_i = 1 + \frac{[2(1-\alpha_{01}) - \rho_i]^2}{4\alpha_{01}\alpha_{10}}, \quad i = 1, 2.$$

Recurrent equations for

$$P_0(t) = F_0(1, 0, 0; t) \tag{6.33}$$

are obtained from the system (6.25)–(6.27) using the substitutions $s_0 = 1$, and $s_1 = s_2 = 0$.

6.1.6 *The single-stage models versus data*

The question considered in this section is whether the single-stage models can simultaneously reproduce the r and P_0 values obtained from experimental data. Each single-stage model yields, for a given value of mutation rate a or mutation probability α and for a given sample size $N(t)$, a uniquely determined pair of values $r(t)$ and $P_0(t)$. We will call the $r - P_0$ plot the set of all such points in the $r - P_0$ plane. The equation of the $r - P_0$ plot can be found by eliminating a (or α) from the expressions for $r(t)$ and $P_0(t)$ in Table 6.2. For example, the $r - P_0$

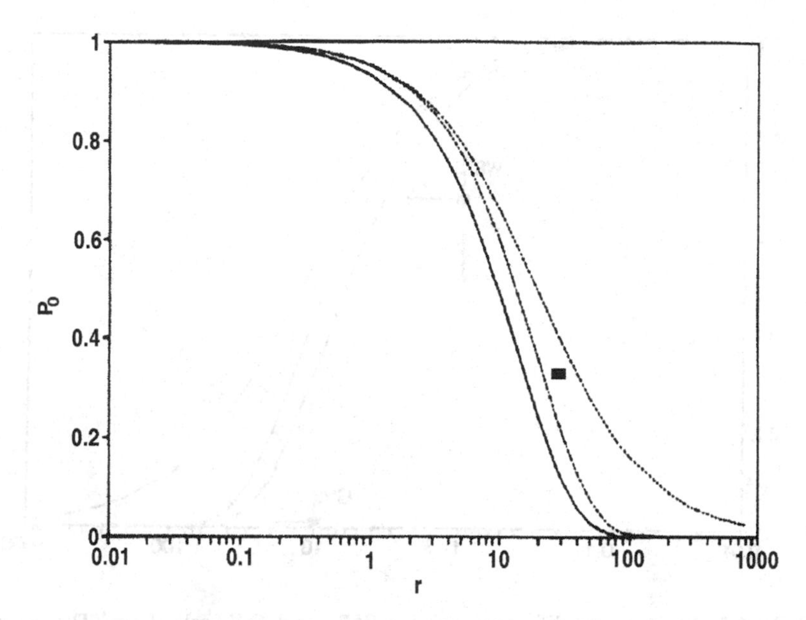

FIGURE 6.2. Bacterial phage resistance data from Experiment 23 of (Luria and Delbrück 1943) and the $r - P_0$ plots ($N = 2.4 \times 10^8$) of the Galton–Watson, Luria–Delbrück, and Markov branching process models. Source: Kimmel, M. and D. Axelrod. 1994. Fluctuation test for two-stage mutations: application to gene amplification. Mutation Research 306: 45–60. Figure 2, page 51. Copyright: 1994 Elsevier Science B.V.

plot of the Luria–Delbrück model has the following equation:

$$P_0 = \exp\left(\frac{-r}{\ln(N)}\right).$$

By graphing the experimentally obtained estimates of r and P_0 together with the corresponding $r - P_0$ plot for an appropriate N, we can verify whether the model can fit the data.

The first series of comparisons includes the original data on bacterial resistance to phage from Luria and Delbrück (1943), almost perfectly matched by the single-stage models. In Figure 6.2 we present the $r - P_0$ plot and the data point of experiment 23 of Luria and Delbrück (1943). Note that the data point interpolates between the models with constant lifetime and exponentially distributed lifetime, two extreme alternatives of lifetime distributions. Experiment 22 provides a similar match. For these classical data, the single-stage models seem perfectly satisfactory.

As a contrast, we analyze the gene amplification data from Tlsty et al. (1989) and Murnane and Yezzi (1988) [details in Kimmel and Axelrod (1994)].

Figures 6.3 and 6.4 demonstrate that the r and P_0 values obtained in this way are not matched by the r–P_0 plots of the Luria–Delbrück, Galton–Watson, and Markov branching process models (from left to right). The overall tendency of these three models is to overestimate either r or P_0. Taking into account cell death makes the match even worse (Fig. 6.5).

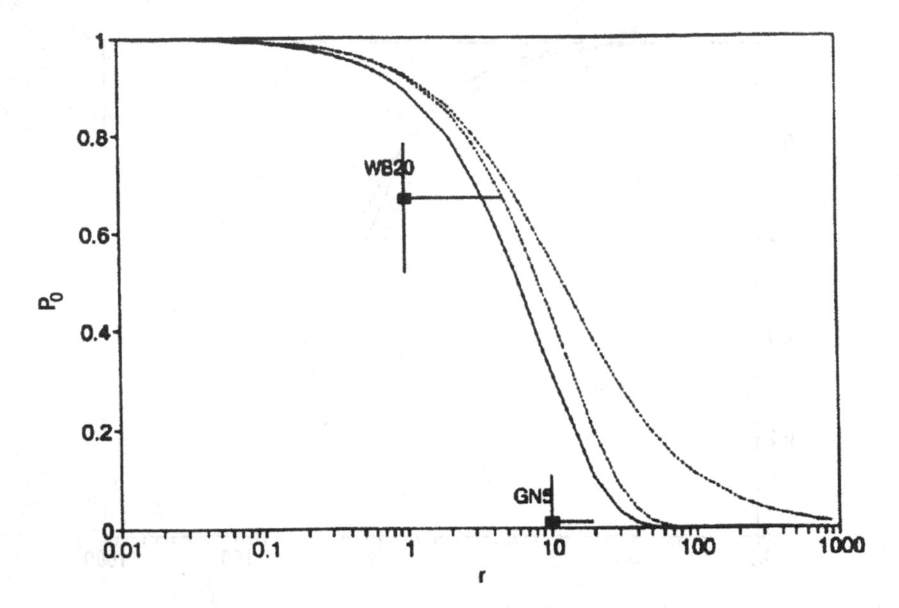

FIGURE 6.3. The gene amplification data for WB_{20} and GN_5 cells from (Tlsty et al. (1989) and the $r - P_0$ plots ($N = 2 \times 10^5$) of the Galton–Watson, Luria–Delbrück, and Markov branching process models (from left to right). Source: Kimmel M. and D. Axelrod. 1994. Fluctuation test for two-stage mutations: application to gene amplification. Mutation Research 306: 45–60. Figure 3, page 52. Copyright: 1994 Elsevier Science B.V.

Figures 6.6 and 6.7 show data on drug resistance due to the loss of HGPRT enzyme activity with the r-P_0 plots. Figure 6.6 includes the data of Morrow (1970) and Figure 6.7 shows the data of Varshaver et al. (1983). Figures 6.6 and 6.7 demonstrate that the r and P_0 values obtained in this way are not matched by the r–P_0 plots of the Luria–Delbrück, Galton–Watson, and Markov branching process models (from left to right).

To visualize the extent of separation of data from the single-stage models, we carried out confidence interval analysis of data. The results are depicted in Figures 6.3–6.7. The vertical error bars are the exact 0.95 confidence intervals for P_0 based on binomial distribution and corrected for plating efficiency. It is difficult to carry out exact analysis for r because its distribution is complicated. Therefore, we only plotted horizontal bars, the right ends of which correspond to the upper 0.95 quantile of the sample. This analysis shows systematic departures from the single-stage model.

6.1.7 The two-stage model versus data

Kimmel and Axelrod (1994) demonstrated that the two-stage model better explains the experimental data concerning drug resistance. The typical estimates of the first-stage forward mutation rate are $\alpha_{01} \approx 10^{-6}$. The corresponding reversal rates are $\alpha_{10} \approx 0.2$–0.95. Finally, the second-step forward mutation rates are

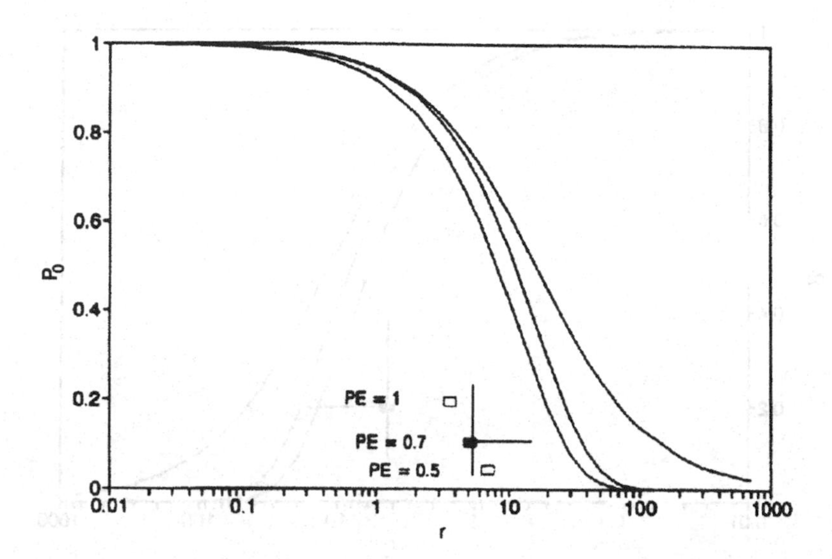

FIGURE 6.4. The gene amplification data for LM205 cells from Murnane and Yezzi (1988) and the $r - P_0$ plots ($N = 1.1 \times 10^7$) of the Galton–Watson, Luria–Delbrück, and Markov branching process models. The three experimental points were obtained using three different hypothetical values of plating efficiency. Source: Kimmel M. and D. Axelrod. 1994. Fluctuation test for two-stage mutations: application to gene amplification. Mutation Research 306: 45–60. Figure 4, page 52. Copyright: 1994 Elsevier Science B.V.

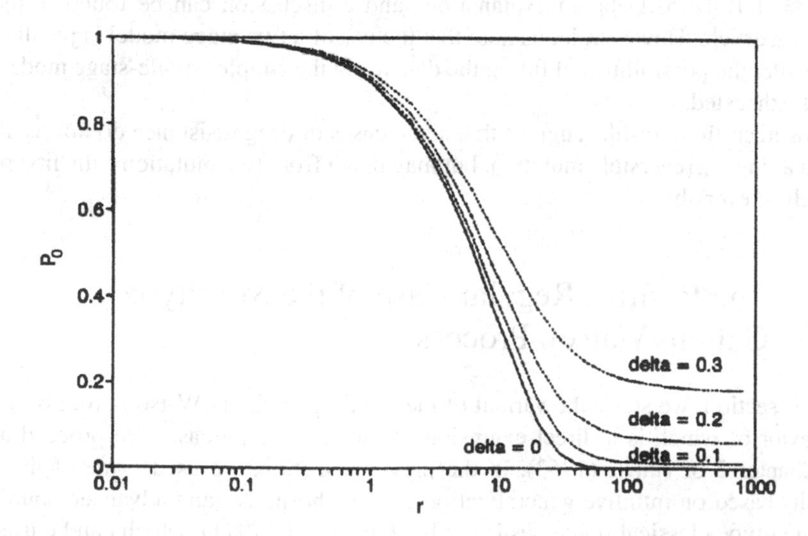

FIGURE 6.5. The effect of cell death on the $r - P_0$ plots of the Galton–Watson model with cell death ($N = 2 \times 10^5$). Delta is the probability of cell death. Source: Kimmel, M. and D. Axelrod. 1994. Fluctuation test for two-stage mutations: application to gene amplification. Mutation Research 306: 45–60. Figure 7, page 52. Copyright: 1994 Elsevier Science B.V.

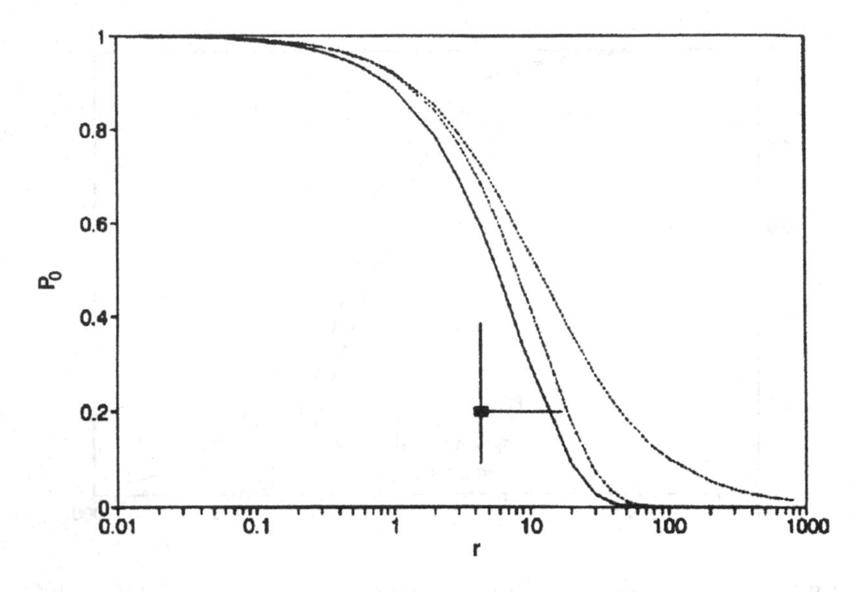

FIGURE 6.6. The drug resistance data for $M - Mc$ mouse cells from experiment 1 of Morrow (1970) and the $r - P_0$ plots ($N = 10^5$) of the Galton–Watson, Luria–Delbrück, and Markov branching process models (from left to right). Source: Kimmel, M. and D. Axelrod. 1994. Fluctuation test for two-stage mutations: application to gene amplification. Mutation Research 306: 45–60. Figure 5, page 52. Copyright: 1994 Elsevier Science B.V.

$\alpha_{12} \approx 0.01$–0.15. Detailed explanations and a discussion can be found in the original article. However, let us note that the use of a two-stage model is justified only after the possibilities of fitting the data using the simpler single-stage models were exhausted.

Together, these results suggest that some cases of drug resistance do not result from a single irreversible mutation, but may result from two mutations, the first of which is reversible.

6.2 The Positive Regular Case of the Multitype Galton–Watson Process

In this section, we study the variant of the multitype Galton–Watson process, the behavior of which is a direct extension of the single-type case. We proceed as in Chapter 2 of Harris (1963). In the previous section, we used some of these results based on intuitive generalizations. An authoritative and advanced source on multitype classical processes is the book by Mode (1971), which can be used as a reference for most of this chapter.

We follow evolution of a population composed of particles of k types. An ancestral particle of type i lives for a unit time interval, and in the moment of death, it produces a random number of progeny particles of generally all k type. The

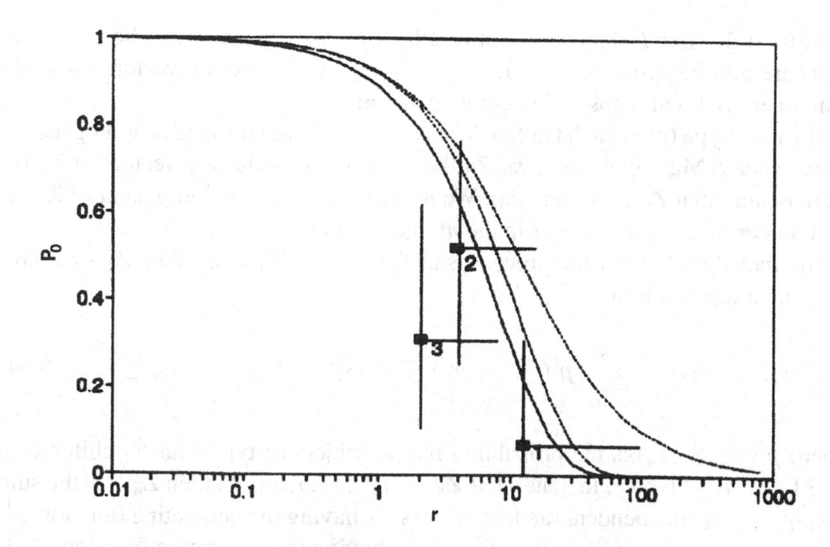

FIGURE 6.7. The drug resistance data for Chinese hamster 237 – 4 cells from replicate cultures 1, 2, and 3, HPRT$^-$ (Varshaver et al. 1983) and the $r - P_0$ plots ($N = 10^5$) of the Galton–Watson, Luria–Delbrück, and Markov branching process models (from left to right). Source: Kimmel, M. and D. Axelrod. 1994. Fluctuation test for two-stage mutations: application to gene amplification. Mutation Research 306: 45–60. Figure 6, page 52. Copyright: 1994 Elsevier Science B.V.

numbers of its progeny of all types constitute a random vector with non-negative integer entries, characterized by pgf $f^i(s_1, \ldots, s_k)$. A progeny of type j starts, independently of all other progeny, a subprocess with itself as the ancestor, by producing at the moment of death, a random vector of progeny of all types, characterized by pgf $f^j(s_1, \ldots, s_k)$. The distribution of this subprocess depends only on the type of the ancestral particle.

The counts of particles of all types existing at time n in the process started by an ancestor of a fixed type constitute a random vector denoted $\mathbf{Z}_n = (Z_n^1, \ldots, Z_n^k)$. The distribution of this vector depends on the type of the ancestral particle of the process. We provide a definition and a theorem stating that in the multitype process, the pgf's of $\mathbf{Z}_n$ are functional iterates of the progeny pgf's, as it was the case for the single-type Galton–Watson process.

6.2.1 *Basics*

The following definition uses a forward approach to the process by relating the numbers of particles in generation $n + 1$ to those in the preceding generation, n. In this way, it underscores the Markov character of the multitype Galton–Watson process.

Definition 5. Let T denote the set of all k-dimensional vectors whose components are non-negative integers. Let $\mathbf{e}_i$, $1 \le i \le k$, denote the vector whose ith component is 1 and whose other components are 0.

The multitype (or vector) Galton–Watson process is a temporally homogeneous vector-valued Markov process $\mathbf{Z}_0, \mathbf{Z}_1, \mathbf{Z}_2, \ldots$, whose states are vectors in T. We shall assume that $\mathbf{Z}_0$ is nonrandom. We interpret Z_n^i, the ith component of $\mathbf{Z}_n$, as the number of objects of type i in the nth generation.

The transition law for the process is as follows. If $\mathbf{Z}_0 = \mathbf{e}_i$, then $\mathbf{Z}_1$ will have the generating function

$$f^i(s_1, \ldots, s_k) = \sum_{r_1,\ldots,r_k}^{\infty} p^i(r_1, \ldots, r_k)\, s_1^{r_1} \cdots s_k^{r_k}, \quad |s_1|, \ldots, |s_k| \le 1, \tag{6.34}$$

where $p^i(r_1, \ldots, r_k)$ is the probability that an object of type i has r_1 children of type $1, \ldots, r_k$ of type k. In general, if $\mathbf{Z}_n = (r_1, \ldots, r_k) \in T$, then $\mathbf{Z}_{n+1}$ is the sum of $r_1 + \cdots + r_k$ independent random vectors, r_1 having the generating function f^1, r_2 having the generating function $f^2, \ldots, r_k$ having the generating function f^k. If $\mathbf{Z}_n = 0$, then $\mathbf{Z}_{n+1} = 0$.

The generating function of $\mathbf{Z}_n$, when $\mathbf{Z}_0 = \mathbf{e}_i$, will be denoted by $f_n^i(s_1, \ldots, s_k) = f_n^i(\mathbf{s})$ $i = 1, \ldots, k$ $n = 0, 1, \ldots$. Then, f_1^i is the function f^i of Eq. (6.34). The vector $(f_n^1(\mathbf{s}), \ldots, f_n^k(\mathbf{s}))$ will be frequently denoted by $\mathbf{f}_n(\mathbf{s})$.

Directly from this definition, we can deduce the following theorem. We omit the details, as they are an extension of those in Section 3.1.2. They can be obtained by a direct application of Theorem 30, part 6 (see Appendix A).

Theorem 21. *The generating functions f_n^i are functional iterates, defined by the relations*

$$\begin{aligned} f_{n+1}^i(\mathbf{s}) &= f^i[f_n^1(\mathbf{s}), \ldots, f_n^k(\mathbf{s})],\ n = 0, 1, \ldots, \\ f_n^0(\mathbf{s}) &= s_i,\ i = 1, 2, \ldots, k. \end{aligned} \tag{6.35}$$

More generally, we have, in vector form

$$\mathbf{f}_{n+N}(\mathbf{s}) = \mathbf{f}_n[\mathbf{f}_N(\mathbf{s})], \quad n, N = 0, 1, 2, \ldots. \tag{6.36}$$

We define $\mathbf{M} = (m_{ij})$ to be the matrix of expected numbers of progeny of all types of parent particles of all types. Specifically, $m_{ij} = \mathrm{E}(Z_1^j|\mathbf{Z}_0 = \mathbf{e}_i) = \partial f^i(1, \ldots, 1)/\partial s_j$, $i, j = 1, \ldots, k$, is the expected number of progeny of type j of a particle of type i. It is assumed that all the first moments m_{ij} are finite and not all equal to 0. By using the chain rule in (6.35), we obtain $\mathrm{E}(\mathbf{Z}_{n+1}|\mathbf{Z}_n) = \mathbf{Z}_n\mathbf{M}$. More generally,

$$\mathrm{E}(\mathbf{Z}_{n+N}|\mathbf{Z}_N) = \mathbf{Z}_N\mathbf{M}^n. \tag{6.37}$$

Analogous expressions for variances are more complicated (see Harris 1963, Mode 1971).

6.2.2 Positivity properties

The following are the essentials of the Perron–Frobenius theory of positive matrices. This theory demonstrates that iterates of positively regular non–negative matrices can be approximated using the powers of the dominating eigenvalue of the matrix, which is shown to be positive. As a consequence, the asymptotic properties of the multitype Galton–Watson process in the positive regular case can be expressed using powers of this eigenvelue.

We shall call a vector or a matrix positive, non-negative, or 0 if all of its components have these properties. If $\mathbf{u}$ and $\mathbf{v}$ are vectors or matrices, then $\mathbf{u} > \mathbf{v}$ ($\mathbf{u} \geq \mathbf{v}$) means that $\mathbf{u} - \mathbf{v}$ is positive (non-negative). Absolute value signs enclosing a vector or a matrix denote the sum of the absolute values of the elements (e.g., $|\mathbf{Z}_n| = \sum_i |Z_n^i|$).

Theorem 22. *Let $\mathbf{M}$ be a non-negative matrix of order k, which is irreducible (i.e., such that $\mathbf{M}^N$ is positive for some positive integer N). Then, $\mathbf{M}$ has a positive eigenvalue ρ that is simple and greater in absolute value than any other eigenvalue; ρ corresponds to positive right and left eigenvectors $\boldsymbol{\mu} = (\mu^i)$ and $\boldsymbol{\nu} = (\nu^i)$, respectively, which are the only non-negative eigenvectors. Moreover, we have*

$$\mathbf{M}^n = \rho^n \mathbf{M}_1 + \mathbf{M}_2^n, \quad n = 1, 2, \ldots, \tag{6.38}$$

where $\mathbf{M}_1 = (\mu^i \nu^j)$, with the normalization $\sum_i \mu^i \nu^i = 1$. Hence, $\mathbf{M}_1 \mathbf{M}_1 = \mathbf{M}_1$. Furthermore:

1. $\mathbf{M}_1 \mathbf{M}_2 = \mathbf{M}_2 \mathbf{M}_1 = 0$.
2. $|\mathbf{M}_2^n| = O(\alpha^n)$ *for some* $\alpha \in (0, \rho)$.
3. *If j is a positive integer, then ρ^j corresponds to $\mathbf{M}^j$ in the same manner as ρ corresponds to $\mathbf{M}$.*

A multitype Galton–Watson process is called positively regular or irreducible if $\mathbf{M}^n$ is positive for some positive integer N.

6.2.3 Asymptotic behavior in the supercritical case

The following result is a direct extension of the analogous result for the single-type process (Theorems 5 and 6).

Theorem 23. *Suppose that the process is positively regular with $\rho > 1$ and that all the second moments of progeny distributions are finite. Then, the random vectors $\mathbf{Z}_n / \rho^n$ converge with probability 1 to a random vector $\mathbf{W}$. Vector $\mathbf{W}$ is nonzero except for trivial cases of all covariance matrices $\mathbf{V}_i = \mathrm{Cov}\,(\mathbf{Z}_1 | \mathbf{Z}_0 = \mathbf{e}_i)$ being zero or $\mathbf{Z}_0 = \mathbf{0}$. If $\mathbf{W}$ is nonzero, then with probability 1 its direction coincides with that of $\boldsymbol{\nu}$, the left eigenvector of $\mathbf{M}$.*

One of the consequences of the theorem is that the limit law in the positively regular case is strictly one dimensional. Although the total number of particles is subject to wide dispersion, their proportions become constant with probability 1.

6.2.4 *Probability of extinction*

It is understandable that the probability of extinction of a multitype process depends on the type of its ancestral particle. Otherwise, the rule is analogous as in the single-type case (Section 3.3). Let q^i be the extinction probability if initially there is one object of type $i = 1, 2, \ldots, k$; that is, $q^i = \mathrm{P}\{\mathbf{Z}_n = 0 \text{ for some } n | \mathbf{Z}_0 = \mathbf{e}_i\}$. The vector $(q^1, \ldots, q^k)$ is denoted by $\mathbf{q}$.

Theorem 24. *Suppose that the process is positively regular and not singular (which would mean that each object has exactly one progeny). If* $\rho \leq 1$, *then* $\mathbf{q} = \mathbf{1}$. *If* $\rho > 1$, *then* $\mathbf{0} \leq \mathbf{q} < \mathbf{1}$ *and* $\mathbf{q}$ *satisfies the equation*

$$\mathbf{q} = \mathbf{f}(\mathbf{q}). \tag{6.39}$$

6.3 *Application*: A Model of Two Cell Populations

The example we present is a simplified version of the model considered in Kimmel and Arino (1991). It is motivated by an experiment described in Sennerstam and Strömberg (1984).

Let us consider two cell populations evolving according to the following rules:

1. Both populations have fixed interdivision times equal to 1.
2. In both populations, the divisions are entirely effective (i.e., each parent cell produces exactly two progeny initially of the same type).
3. After division each type-1 progeny (independent of the other) switches to type 2 with probability p_{12} and remains type 1 with probability $p_{11} = 1 - p_{12}$.
4. Analogously, each type-2 progeny (independent of the other) switches to type 1 with probability p_{21} and remains type 2 with probability $p_{22} = 1 - p_{21}$.

The known biological example is the population of cultured transformed embryonic cells maintained by Sennerstam. The "normal" embryonic cells have a program to switch irreversibly from one developmental stage to the next. The transformed cells are maintained indefinitely because they switch back and forth between two stages, named 1 and 2 by us. Under the simplified assumptions specified above, their proliferation is described by a 2-type Galton–Watson process.

The progeny pgf's of the process are

$$f^1(s_1, s_2) = (p_{11}s_1 + p_{12}s_2)^2, \tag{6.40}$$

$$f^2(s_1, s_2) = (p_{21}s_1 + p_{22}s_2)^2. \tag{6.41}$$

The expected progeny matrix of the process is equal to

$$\mathbf{M} = \begin{pmatrix} 2p_{11} & 2p_{12} \\ 2p_{21} & 2p_{22} \end{pmatrix}. \tag{6.42}$$

The eigenvalues of matrix $\mathbf{M}$ are found from the equation

$$\rho^2 - \rho(2p_{11} + 2p_{22}) + 4(p_{11}p_{22} - p_{12}p_{21}) = 0.$$

The greater of the two real roots of this equation (the Perron–Frobenius root or eigenvalue) is equal to $\rho = 2$. The left eigenvector $\boldsymbol{\nu}$ corresponding to the Perron–Frobenius root is the row vector satisfying the matrix equation $\nu(\mathbf{M} - 2\mathbf{Id}) = \mathbf{0}$, or

$$2(\nu_1, \nu_2)\begin{pmatrix} -p_{12} & p_{12} \\ p_{21} & -p_{21} \end{pmatrix} = \mathbf{0}. \tag{6.43}$$

We obtain

$$\frac{\nu_1}{\nu_2} = \frac{p_{21}}{p_{12}}.$$

The process is positively regular. Theorem 23 yields that with probability 1,

$$(Z_n^1, Z_n^2) \sim 2^n(\nu_1, \nu_2)W, \quad n \to \infty,$$

where W is a scalar random variable.

The meaning of this result is that the proportion of the type-1 and type-2 cells is asymptotically determined by the ratio $\nu_1/\nu_2 = p_{21}/p_{12}$. The interesting feature is that both p_{21} and p_{12} can be very small (i.e., that the switching between both types is not frequent), and still the proportion is maintained. For the experimental data, it was estimated that p_{12} and p_{21} are of the order of 0.1 (Arino and Kimmel 1991).

6.4 *Application*: Stochastic Model of the Cell Cycle with Chemotherapy

The current application does not draw on the theory in the previous section. Instead, it is an example of a model using a multitype Bellman–Harris process.

The goal of cancer chemotherapy is to stop tumor cells from dividing and to kill them while sparing normal cells. Some chemotherapy protocols depend on the differential effect of drugs on cells in different compartments of the cell cycle. For instance, combination drug chemotherapy may utilize two drugs which affect cells in different compartments of the cell cycle with different efficiencies. Such combination chemotherapy is expected to be more effective in tumor cell populations than in normal cell populations. The rationale is that tumor cell populations have a larger fraction of cells progressing through the cell cycle than normal cells. This approach requires knowledge of the "drug action curve," the percentage of cells affected depending on their position in the cell cycle. In Section 5.4, we developed a method of estimating the duration of cell cycle compartments, based on stathmokinetic experiments. This method is now extended to determine the relative effects of a drug on cells in different compartments of the cell cycle.

Modern technology allows the determination of the amount of DNA per cell by measuring the fluorescence of stained DNA excited by a laser in a flow cytometer. This has lead to an improved stathmokinetic method that utilizes the amount of DNA per cell, rather than the number of cells in mitosis, as a function of time for which the cells are exposed to the statmokinetic agent. The means and variances of durations of each of the cell cycle compartments can then be estimated using the mathematical methods described in Section 5.4.

Additional mathematical methods are required to obtain the estimates of the cell-cycle-specific effects of anticancer drugs. We develop a model which describes the flow of cells through successive compartments of the cell cycle. The model allows the estimation of the fraction of cells blocked in each cell cycle compartment by an anticancer agent.

This application is mainly based on the article by Kimmel and Traganos (1986). It is the continuation of the stathmokinetic analysis example of Section 5.4. The mathematical tool we use is the multitype Bellman–Harris process. We do not develop a rigorous theory, but employ intuition and analogies with the Galton–Watson branching process. For a related, more mathematical approach, see Crump (1970).

We want to model the long-term in vitro effects of an anticancer drug, acting with a different strength on cells in different phases of the cell cycle, based on the short-term observations collected using the stathmokinetic experiment. For this purpose, we decompose the cell cycle into a sequence of compartments differing with respect to sensitivity to the drug. These compartments may be different from individual cell cycle phases. Specifically, in the current model, the S phase is subdivided into a number of smaller compartments, to account for different sensitivities of cells in different stages of DNA synthesis.

6.4.1 Model of drug-perturbed stathmokinesis

The following model is employed to analyze the drug action (Fig. 6.8): The cell cycle is divided into M disjoint compartments. The cell residence time in the mth compartment is an independent random variable with distribution density $p_m(\cdot)$. The conditions of the stathmokinetic experiment are satisfied, by assuming that there is no cell inflow into the first compartment nor cell outflow from the last (Mth) compartment. In each compartment, exposure to a given concentration of the drug causes a permanent block for a fraction $1 - u_m$ of cells, which would otherwise leave this compartment. By choosing a sufficiently dense subdivision of the cell cycle into compartments, it is possible to construct a curve of drug action, the coordinates of which are the quantities $1 - u_m$.

Let us denote by $N_m(t)$ the expected cell count in the mth compartment and by $x_m(t)$ the expected cell outflow rate from the mth compartment, at time t. We have

$$\begin{aligned} N_1(t) &= N_1(0) - \int_0^t x_1(s)\,ds, \\ N_m(t) &= N_m(0) + \int_0^t [x_{m-1}(s) - x_m(s)]\,ds, \quad m = 2, \ldots, M-1, \qquad (6.44) \\ N_M(t) &= N_M(0) + \int_0^t x_{M-1}(s)\,ds. \end{aligned}$$

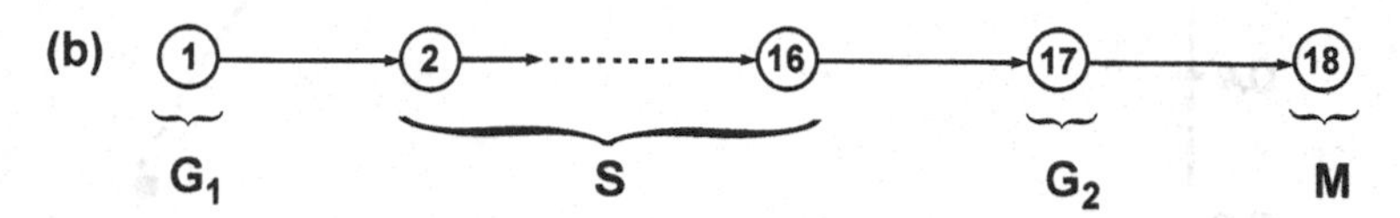

FIGURE 6.8. (a) The model of blocking drug action. The cell cycle is divided into M compartments. There is no cell flow into the first compartment, nor cell outflow from the last compartment. Notation: $p_m(t)$, distribution density of the residence time; $x_m(t)$, outflow rate; $N_m(t)$, cell count: u_m, cell fraction in the mth compartment not blocked by the drug. (b) Correspondence between compartment number and cell cycle phase. Source: Kimmel, M. and Tranganos, F. 1986. Estimation and prediction of cell cycle specific effects of anticancer drugs. Mathematical Biosciences 80: 187–208. Figure 1, page 191. Copyright: 1986 Elsevier Science Publishing Co., Inc.

It is assumed that before the beginning of stathmokinesis (i.e., for $t < 0$), the cell population was in the exponential steady state (ESS) (i.e., expected cell counts in all the cell cycle compartments were proportional to e^{bt}). The constant b is the Malthusian parameter of exponential growth.

Balancing of expected ESS cell flows from one cell cycle compartment to another, as described in more detail in Kimmel (1980 a, 1980 b), we obtain

$$\begin{aligned} N_1(0) &= 2(1 - \hat{p}_1), \\ N_m(0) &= 2\hat{p}_1 \cdots \hat{p}_{m-1}(1 - \hat{p}_m), \end{aligned} \tag{6.45}$$

where $\hat{p}_m$ is the Laplace transform of the distribution $p_m(\cdot)$, evaluated at b:

$$\hat{p}_m = \int_0^\infty p_m(t)e^{-bt}\, dt. \tag{6.46}$$

Computation of the outflows $x_m(\cdot)$ perturbed by the drug is more complicated. Except for $x_1(\cdot)$, the cell outflow is the sum of a component from the outflow of the preceding compartment and another component from the initial distribution (at $t = 0$) of cells in this compartment:

$$x_m(t) = u_m\left[x_{m-1}(t) * p_m(t) + x_m^0(t)\right], \quad m = 2, ..., M-1, \tag{6.47}$$

where the asterisk denotes the convolution of functions $(f * g)(t) = \int_0^t f(t - \tau)g(\tau)\, d\tau$. The flow $x_m^0(t)$ can be calculated in the following way: Let us denote by $p_{1m}(t)$ the distribution of the sum of residence times in compartments 1 through m, and by $P_{1m}(t)$ the corresponding cumulative distribution. Also, let us denote

$$a_{1m}(t) = e^{bt}\int_t^\infty p_{1m}(s)e^{-bs}\, ds = e^{bt}\hat{p}_{1m} - e^{bt} * p_{1m}(t), \tag{6.48}$$

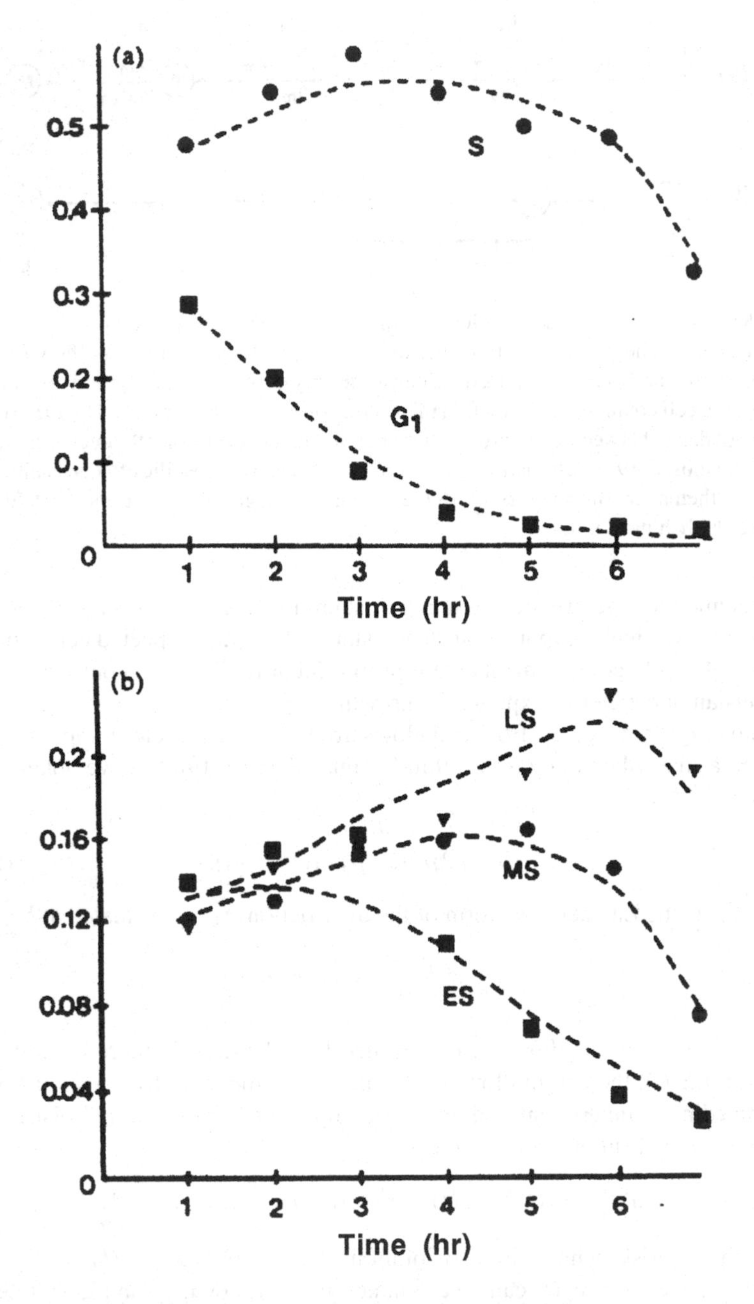

FIGURE 6.9. Stathmokinetic data (low drug concentration) fitted by the model curves: (a) S (circles) and G_1 (squares) phases, (b) early (channels 27–30, squares), mid (channels 32–35, circles), and late (channels 37–40, triangles) S-phase "windows." Source: Kimmel, M. and Tranganos, F. 1986. Estimation and prediction of cell cycle specific effects of anticancer drugs. Mathematical Biosciences 80: 187–208. Figure 4, page 198. Copyright: 1986 Elsevier Science Publishing Co., Inc.

$$a_m(t) = e^{bt} \int_t^\infty p_m(s) e^{-bs}\, ds = e^{bt} \hat{p}_m - e^{bt} * p_m(t). \tag{6.49}$$

We have $\hat{p}_{1m} = a_{1m}(0)$ and $\hat{p}_m = a_m(0)$. In Proposition 1, we found the asymptotics of the number of cells in phase 1 of the cell cycle, when the cell cycle is subdivided into two phases, under normal conditions in a stathmokinetic experiment not perturbed by any other agent. We can consider our compartments 1 through m as a phase 1, and by doing so, we obtain, by Proposition 1,

$$\bar{N}_{1m}(t) = 2\left[1 - P_{1m}(t) - a_{1m}(t)\right]. \tag{6.50}$$

Let us note that, by Eq. (6.48), we have $d[a_{1m}(t)]/dt = ba_{1m}(t) - p_{1m}(t)$, which implies

$$\frac{d[\bar{N}_{1m}(t)]}{dt} = -2ba_{1m}(t). \tag{6.51}$$

The outflow $x_m^0(t)$ from the initial distribution of cells in compartment m is the same whether or not a perturbing agent (other than the stathmokinetic agent) is applied. It is equal to the total outflow from compartments 1 through m, minus a component due to the outflow from compartments 1 through $m-1$:

$$\begin{aligned} x_m^0(t) &= \frac{d[-\bar{N}_{1m}(t)]}{dt} - \frac{d[-\bar{N}_{1,m-1}(t)]}{dt} * p_m(t) \\ &= 2b[a_{1m}(t) - a_{1,m-1}(t) * p_m(t)] \\ &= 2b\{e^{bt}\hat{p}_{1m} - e^{bt} * p_{1m}(t)] \\ &\quad -[e^{bt}\hat{p}_{1,m-1} - e^{bt} * p_{1,m-1}(t)] * p_m(t)\} \\ &= 2b\{e^{bt}\hat{p}_{1,m-1}\hat{p}_m - e^{bt} * p_{1m}(t)] \\ &\quad -[e^{bt} * p_m(t)\hat{p}_{1,m-1} - e^{bt} * p_{1,m-1}(t) * p_m(t)]\} \\ &= 2b\hat{p}_{1,m-1}a_m(t). \end{aligned} \tag{6.52}$$

Combining Eqs. (6.47) and (6.52), we write the following recurrence:

$$\begin{aligned} x_1(t) &= 2ba_1(t), \\ x_m(t) &= u_m[x_{m-1}(t) * p_m(t) + 2b\hat{p}_{1,m-1}a_m(t)], \quad m = 2, ..., M-1. \end{aligned} \tag{6.53}$$

Based on Eqs. (6.53), an explicit expression is derived:

$$x_m(t) = 2b\left\{\sum_{i=1}^{m}\left(\prod_{j=1}^{i-1}\hat{p}_j\right)\left(\prod_{j=i}^{m}u_j\right)a_i(t) * [p_{i+1}(t) * p_{i+2}(t) * \cdots * p_m(t)]\right\},$$
$$m = 1, ..., M-1. \tag{6.54}$$

6.4.2 Model parameters

It is generally true that the structure of a model depends on the precision of the measurements. In the present case, we divide the cell cycle into smallest compartments in which it is possible to follow the cell count. A fine subdivision is possible in the S phase: We can consider the cells ascending from lower to higher DNA content. Therefore, the model has structure as depicted in Figure 6.8b: Compartment 1 is the G_1 phase, compartments 2–16 cover the S phase, compartment 17 is G_2, and compartment 18 is M.

The main source of variability in the cell's generation time is its transit through the G_1 phase. In practice, the durations of all the other cell cycles phases can be considered nonrandom. The distribution of cell residence time in G_1 was estimated (Fig. 6.10) with the aid of a nonparametric procedure presented in Section 5.4.

The deterministic residence times in the remaining cell cycle compartments were assessed based on the ESS cell counts in these compartments. Their estimation as well as estimation of the coefficients u_m characterizing drug action is described in Kimmel and Traganos (1986).

6.4.3 *Prediction of the effects of continuous exposure to the drug*

Figure 6.11 presents the model used to predict effects of continuous exposure to the drug. It is assumed that once a cell is blocked, it does not progress further through the cell cycle. In the model, the blocked cells pass to the "primed" compartments: G_1', S', or $(G_2 + M)'$. The $(G_2 + M)'$ compartment also contains those G_2 cells which progressed to M but did not divide; instead, they increased their ploidy by, for example, defective cytokinesis.

Simulation of the effects of continuous exposure to the drug based on this model was carried out analogously to similar simulations in Kimmel and Traganos (1985) or in Darzynkiewicz et al. (1984); for a more theoretical treatment, see Kimmel (1980c).

6.4.4 *Results*

The estimates of the basic parameters of the cell cycle of exponentially growing Friend erythroleukemia cells are as follows: The average residence time in G_1, $E(T_{G_1}) = 3.43$ h; the residence time in S, $T_S = 5.08$ h; in G_2, $T_{G_2} = 2.21$ h; in M, $T_M = 0.60$ h; and the growth rate (Malthusian parameter), $b = 0.062\ h^{-1}$, corresponds to the doubling time of 11.22 h.

The fractions of cells blocked by the drug in the cell cycle compartments defined in the previous section were computed from the drug-perturbed stathmokinetic data. They are presented in Figure 6.12, for low (10 n*M*) and high (50 n*M*) concentration of the drug. As evident from this graph, the blocking action of the drug is higher for cells more advanced in their progression through S. The durations (T_j) of the 15 successive subcompartments of the S phase are not very different from each other (mean duration: 0.34 h; coefficient of variation: 0.13)

Fits to the stathmokinetic data obtained using the low-concentration drug action curve of Figure 6.12 are presented in Figure 6.9. For the high drug concentration, the quality of the fit is similar.

Modeling of cell kinetics under continuous exposure to the drug, employing the drug action curves estimated from the stathmokinetic experiment (Fig. 6.12), is

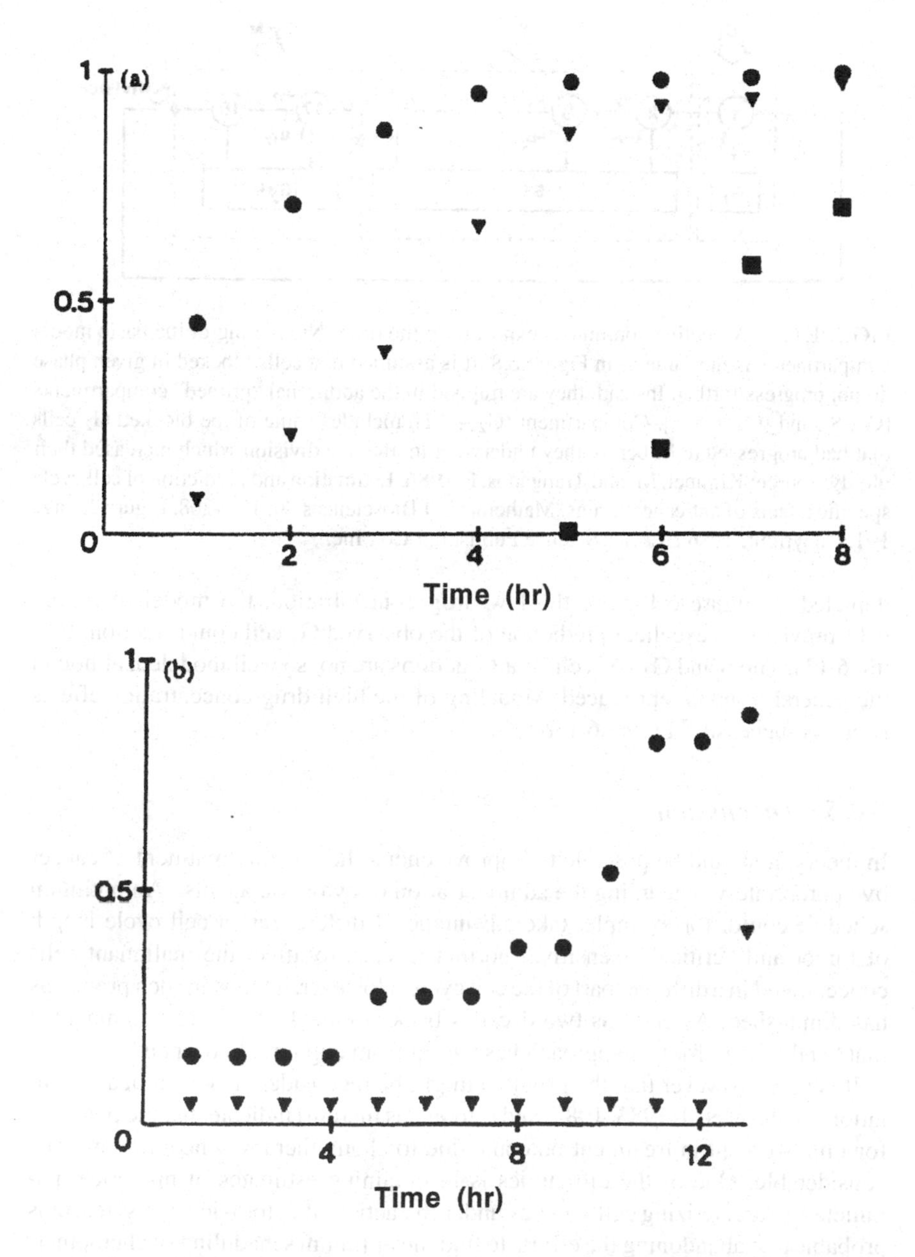

FIGURE 6.10. Example of nonparametric estimation of distribution $P_1(t)$ based on exit curve $f_1(t)$. (a) Friend erytholeukemia cells: circles, G_{1A}; triangles G_1; squares, $G_1 + S$. (b) L-cells: circles, G_1; triangles $G_1 + S$. Source: Kimmel, M. and Tranganos, F. 1986. Estimation and prediction of cell cycle specific effects of anticancer drugs. Mathematical Biosciences 80: 187–208. Figure 8, page 205. Copyright: 1986 Elsevier Science Publishing Co., Inc.

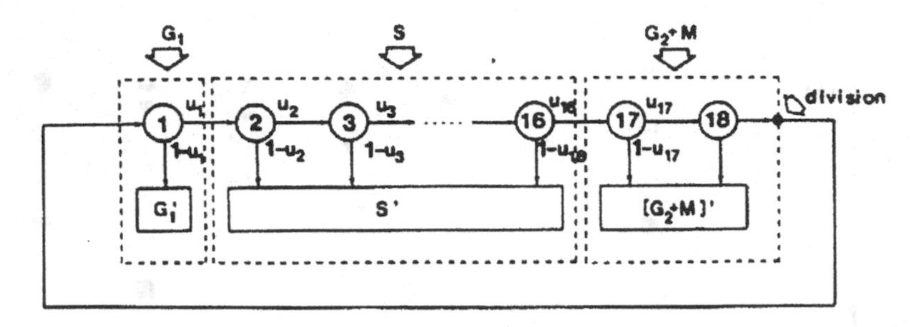

FIGURE 6.11. Modeling continuous exposure to the drug. Numbering of the basic model compartments is the same as in Figure 6.8. It is assumed that cells blocked in given phase do not progress further. Instead, they are trapped in the additional "primed" compartments [G_1', S', and $(G_2 + M)'$]. Compartment $(G_2 + M)'$ includes some of the blocked G_2 cells that had progressed to M before they underwent ineffective division which increased their ploidy. Source: Kimmel, M. and Tranganos, F. 1986. Estimation and prediction of cell cycle specific effects of anticancer drugs. Mathematical Biosciences 80: 187–208. Figure 2, page 191. Copyright: 1986 Elsevier Science Publishing Co., Inc.

depicted in Figure 6.13. For the low drug concentration, the model of Figure 6.11 provides an excellent prediction of the observed G_1 cell count fraction, Figure 6.13a. The S and G_2+M cell count fractions are not so well modeled, although the general trend is reproduced. Modeling of the high-drug-concentration effects is not as successful, Figure 6.13b.

6.4.5 *Discussion*

In theory, it should be possible to improve chemotherapeutic treatment of cancer by appropriately scheduling the administration of cytotoxic agents . An optimum schedule could, for example, take advantage of differences in cell cycle length of tumor and "critical" (sensitive) normal tissues, to affect the malignant cells concentrated in a different part of the cell cycle. However, interest in such proposals has diminished. As early as two decades back, Tannock (1978) has commented that "enthusiasm for this approach has varied from euphoria to despair."

It appears however that the problem might be reconsidered. Theoretical calculations (Dibrov et al. 1983, 1985, and references therein) indicate that the potential for improvement in treatment outcome due to chemotherapy scheduling may be considerable. One of the difficulties is in obtaining estimates of numerical parameters characterizing cell kinetics under the action of cytotoxic agents. It seems probable that abandoning the efforts to find the optimum scheduling of chemotherapy was caused largely by the inability to find good estimates of the parameters mentioned earlier.

The failure to predict effects of the long-term (continuous) exposure to the drug at the higher concentration (see Fig. 6.13b) is probably related to considerable cell damage at this concentration. This damage is not apparent in the course of the stathmokinetic experiment (in fact, the drug action curves for the two drug con-

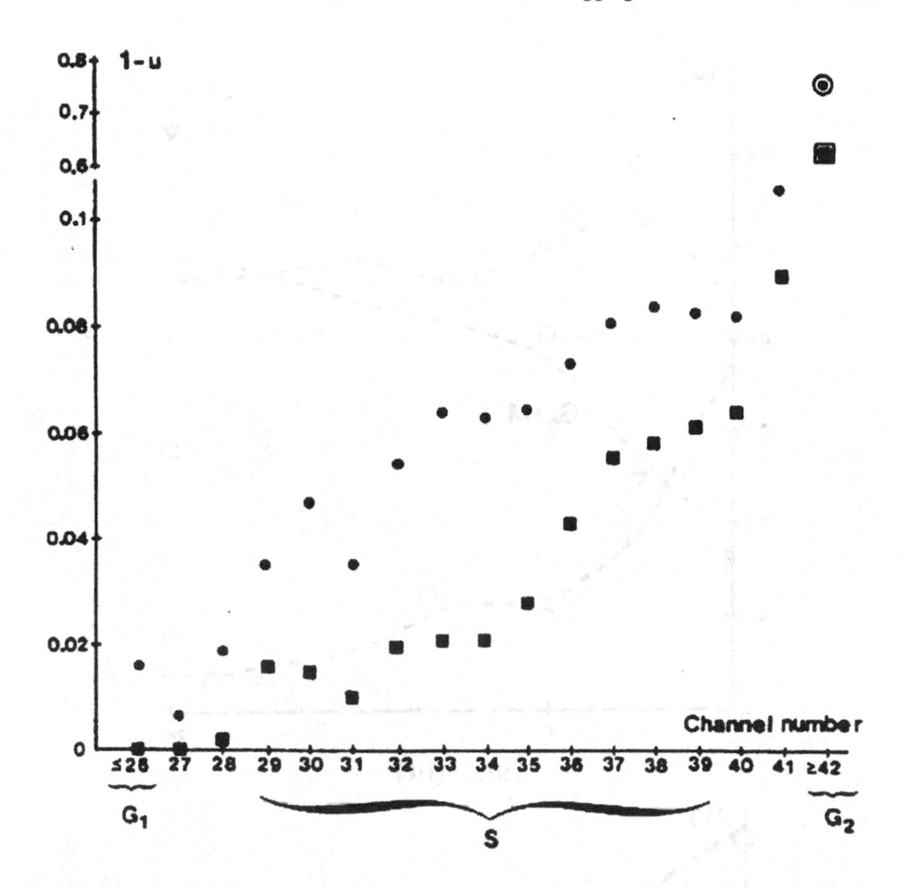

FIGURE 6.12. Drug action curves for the low (10 nM, squares) and high (50 nM, circles) drug concentrations. Fractions of cells blocked by the drug in given cell cycle compartments are plotted against corresponding numbers of the flow cytometer channels. The G_2 fractions (channel ≥ 42) are depicted on a different scale. Source: Kimmel, M. and Tranganos, F. 1986. Estimation and prediction of cell cycle specific effects of anticancer drugs. Mathematical Biosciences 80: 187–208. Figure 3, page 197. Copyright: 1986 Elsevier Science Publishing Co., Inc.

centrations differ only slightly), but it probably manifests itself during subsequent cell divisions.

6.5 *Application*: Cell Surface Aggregation Phenomena

This model is taken from the book by Macken and Perelson (1985). Molecules on the cell surface (receptors) are activated by contact with molecules in the extracellular medium (ligands). The activated receptors initiate signaling pathways within cells resulting in cell proliferation and cell differentiation. The strength of

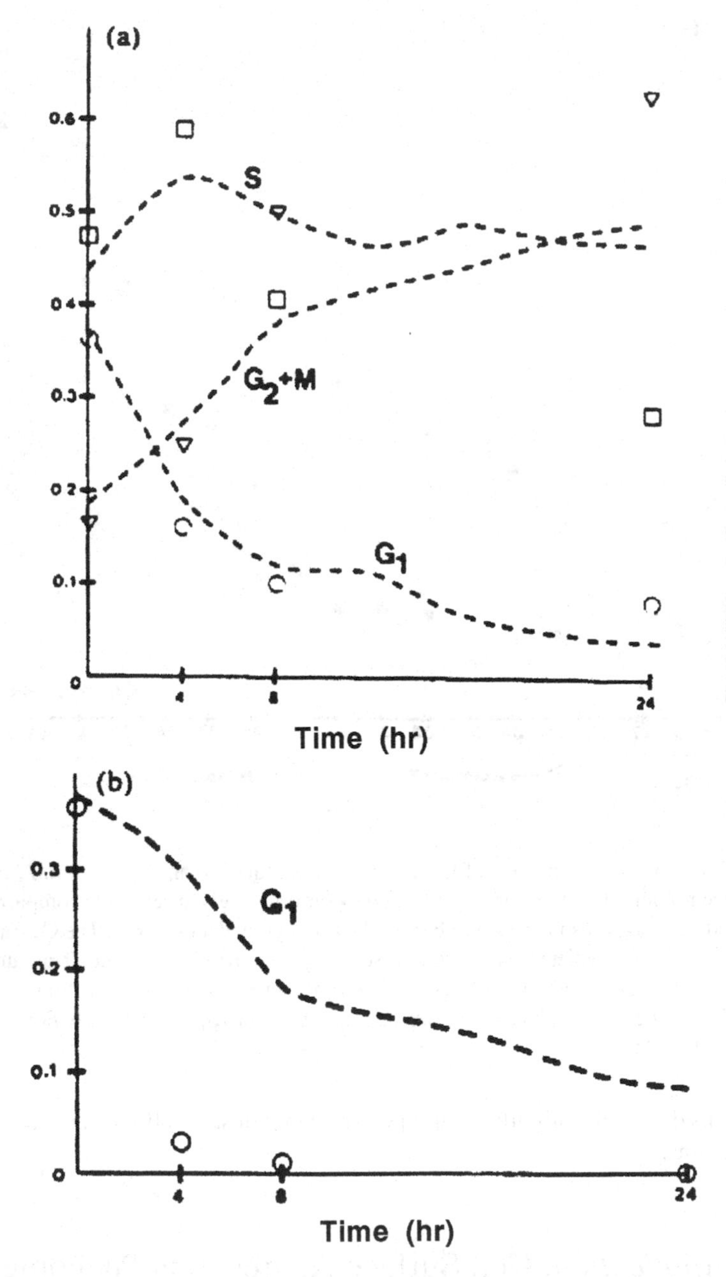

FIGURE 6.13. Observed versus measured cell count fractions in different phases of the cell cycle, under continuous exposure to the drug: (a) low concentration (10 n*M*), (b) high concentration (50 n*M*). Measurements: circles, G_1; squares, S; triangles, $G_2 + M$. Source: Kimmel, M. and Tranganos, F. 1986. Estimation and prediction of cell cycle specific effects of anticancer drugs. Mathematical Biosciences 80: 187–208. Figure 5, page 199. Copyright: 1986 Elsevier Science Publishing Co., Inc.

the signals depends on the specificity of the interaction between a ligand and its receptor, and the number of activated receptors per cell.

Examples of ligands are hormones such as insulin and growth factors and antigens such as proteins on the surface of bacteria and viruses. Some receptors and some ligands are multivalent (e.g., the receptor molecules can react with more than one ligand at a time and the ligands can react with more than one receptor at a time). Multivalency may result in clusters of ligand–receptor complexes. It is of interest to determine the size distribution of the aggregates and the probability that they will continue to increase in size or stop increasing in size.

6.5.1 Relationship between the Galton–Watson process and the aggregation process

Let us suppose for the beginning that we are given a collection of m-valent particles of single type (an m-valent particle is one that can bind m other particles). We restrict our attention to the aggregates of these particles that contain no loops and, hence, have the topological form of a tree. We equate the probability of k particles being bound to a given particle, with p_k, the probability of this particle having k offspring. The particle valency in the aggregation process is accounted for in the Galton–Watson process by imposing a restriction on the maximum possible number of offspring contributed by a single parent to the next generation. Thus, a parent in generation 0 can have at most m offspring, whereas a parent in later generations can have at most $m - 1$ offspring, because one particle site is used to attach the particle to its own parent (Fig. 6.14).

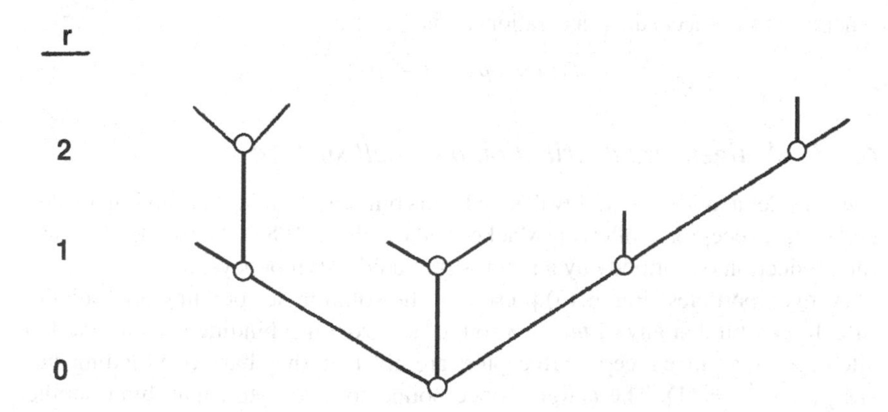

FIGURE 6.14. A typical family tree representing the aggregation of f-valent particles. Here, $f = 3$. Note that particles in generation $n = 0$ can have at most f offspring, whereas in all later generations, a particle can have at most $f - 1$ offspring. Source: Macken, C.A. and Perelson, A.S. 1985. Branching Processes Applied to Cell Surface Aggregation Phenomena. A Multitype Branching Process Model. Lecture Notes in Biomathematics 58. Springer-Verlag. Berlin. Figure 2.2, page 16. Copyright: 1985 Springer-Verlag.

To summarize, the analogy between Galton–Watson process and aggregation processes is that an n-mer is represented by a rooted tree containing n nodes, with the degree of the root being at most m and the degree of all other nodes being at most $m - 1$.

The purpose of the mathematical representation is to find the distribution of the sizes of aggregates. The total size Y of the aggregate is equal to the summary number of objects produced in all generations of the Galton–Watson process:

$$Y = \sum_{n=0}^{\infty} Z_n. \tag{6.55}$$

We are interested in the distribution of random variable Y including cases when Y is infinite. This last possibility corresponds to the so-called gelation in which the aggregation process escapes control and utilizes all the particles suspended in the medium. Let us note that Y can be finite only if the process $\{Z_n, n = 1, 2, \ldots\}$ dies out with probability 1, (i.e., in the subcritical and critical cases). In the supercritical process, there exists the nonzero probability $1 - q$ that the number of generations is infinite. This latter is the probability of gelation.

6.5.2 *Progeny distributions*

We have to specify p_k, the probability that k sites of a randomly chosen m-valent particle are bound. Let p be the probability that a randomly chosen site is bound. Then, because sites act independently, p_k is given by the binomial formula. Consequently, the progeny pgf in the 0th generation is

$$f_0(s) = (ps + 1 - p)^m,$$

whereas in the succeeding generations, it is

$$f(s) = (ps + 1 - p)^{m-1}.$$

6.5.3 *Antigen size distribution on a cell surface*

We consider a model for multivalent antigens binding to and cross-linking bivalent cell surface receptors, following Macken and Perelson (1985). The model describes the production of antibody by antigen-stimulated B-lymphocytes.

Antigen particles (Fig. 6.15) present in the solution surrounding a population of cells can bind at any of $m_a = 3$ (out of six existing) binding sites to one free site of a cell surface receptor. Receptors are bivalent; they have two binding sites (i.e., $m_r - 1 = 1$). The antigen, once bound to a receptor, may bind another single receptor site at any out of remaining $m_a - 1 = 2$ sites or it may bind two free sites of two receptors. In the model, the two antigen sites are not allowed to bind to two sites of a single receptor, as this would violate the tree structure. Repeated binding creates patches of antigen particles cross-linking receptors on the cell surface. Gelation is equivalent to formation of "infinite-size" (very large) antibody-receptor clusters on cell surface.

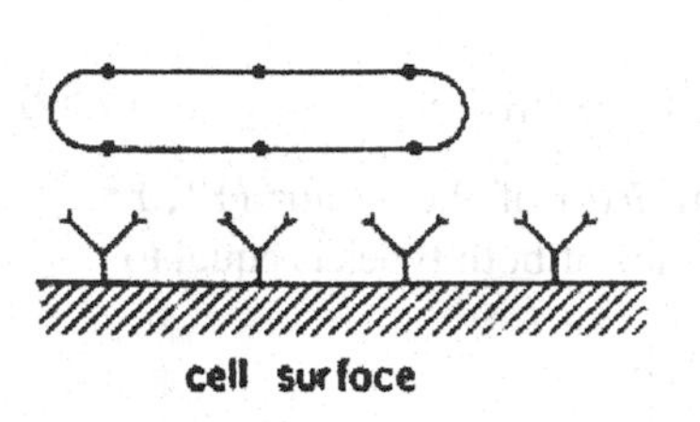

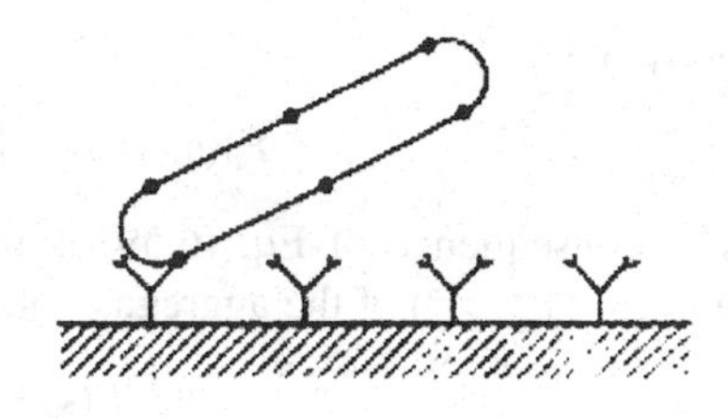

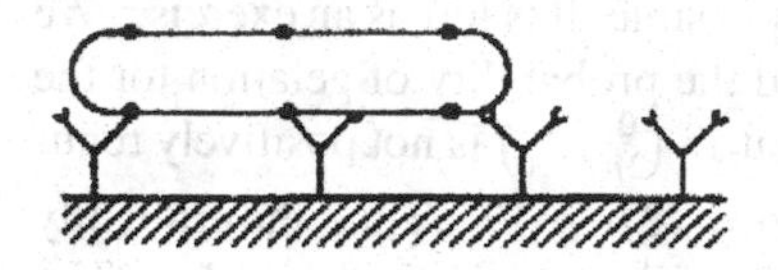

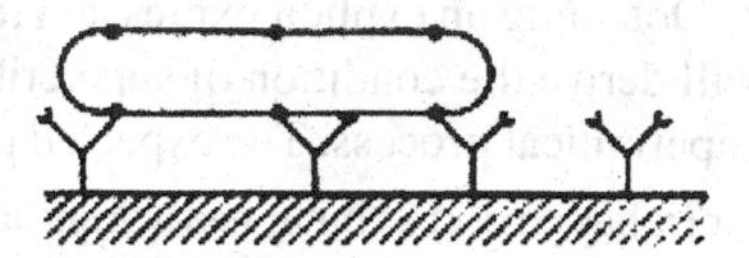

FIGURE 6.15. A model for multivalent antigens binding to and cross-linking bivalent cell surface receptors. The antigen, present in the solution surrounding a population of cells, can bind at any of $f = 3$ (out of six existing) binding sites to one site of a free cell surface receptor. Receptors are bivalent (i.e., have two binding sites). The antigen, once bound to a receptor, may bind another receptor at any out of remaining $f - 1 = 2$ sites and so forth. (Modified from Macken and Perelson, 1985.) Source: Macken, C.A. and Perelson, A.S. 1985. Branching Processes Applied to Cell Surface Aggregation Phenomena. A Multitype Branching Process Model. Lecture Notes in Biomathematics 58. Springer-Verlag. Berlin. Figure 4.2, page 51. Copyright: 1985 Springer-Verlag.

The special type of aggregate described is distinguished by the fact that antigens and antibodies alternate along any path through the aggregate. Consequently, the model is described by a two-type Galton–Watson process $\mathbf{Z}_n = (Z_n^1, Z_n^2)$, $n = 1, 2, \ldots$, in which the offspring of the type-1 object (receptor particle) is of type-2 only and the offspring of the type-2 object (antigen particle) is of type 1 only, that is,

$$f^1(\mathbf{s}) = f^1(s_2) = (p_1 s_2 + 1 - p_1), \tag{6.56}$$

$$f^2(\mathbf{s}) = f^2(s_1) = (p_2 s_1 + 1 - p_2)^2. \tag{6.57}$$

We suppose that the process (aggregate) is started by a single receptor particle and, therefore, for the 0th generation,

$$f_0^1(s_2) = (p_1 s_2 + 1 - p_1)^2. \tag{6.58}$$

Calculations based on Eqs. (6.56)–(6.58) show that the pgf $F_n(\mathbf{s})$ of the vector (Y_n^1, Y_n^2) of the counts of all particles of both types up to generation n,

$$(Y_n^1, Y_n^2) = \sum_{i=0}^{n} (Z_i^1, Z_i^2),$$

is equal to

$$F_n(\mathbf{s}) = s_1 f_0^1\{s_2 f^2[s_1 f^1(s_2 \cdots)]\}. \tag{6.59}$$

The consequence of Eq. (6.59) is that the pgf $F(\mathbf{s})$ of the vector $(Y^1, Y^2) = \lim_{n\to\infty}(Y_n^1, Y_n^2)$ of the aggregate totals of particles of both types is equal to

$$F(\mathbf{s}) = s_1 f_0^1[\Phi(\mathbf{s})],$$

where $\Phi(\mathbf{s})$ is the solution of the equation

$$\Phi(\mathbf{s}) = s_2 f^2\{s_1 f^1[\Phi(\mathbf{s})]\}. \tag{6.60}$$

The pgf solution of Eq. (6.60) always exists because of the monotone convergence. It may correspond to infinite particle count (i.e., gelation), if $\Phi(1, 1) < 1$.

Obtaining an explicit expression for $\Phi(\mathbf{s})$ is possible. It is left as an exercise. We will derive the condition of supercriticality and the probability of gelation for the supercritical process. The expected progeny matrix $\binom{0 \quad p_1}{2p_2 \quad 0}$ is not positively regular (it has period 2) but it has a dominating root $\rho = (2p_1p_2)^{1/2}$. Thus, the criticality parameter is proportional to the geometric mean of the reactivities p_1 and p_2. The probability of gelation is equal to $1-(1-p_1)^2q_2$, where $q_2 = (1-p_1p_2)^2/(p_1p_2)^2$ is obtained by solving equation $(q_1, q_2) = [f^1(q_1, q_2), f^2(q_1, q_2)]$.

The above expressions are valid in the supercritical case.

6.6 Sampling Formulas for the Multitype Galton–Watson Process

The literature on multitype branching processes is mostly focused on asymptotic theory. In comparison, relatively little has been done to address problems of sampling in finite time from a branching process. This is a problem which is relevant in many biological applications. In PCR, the polymerase chain reaction, genetic material is amplified and sampled after a fixed number of cycles. In cell cultures, cells are grown and harvested after a fixed number of population doublings. Also, many branching processes arising in these applications are intrinsically reducible in the sense that some types can only have certain other types in their ancestries. In such processes, limiting distributions on the type space are typically degenerate and of no practical use.

In this section, we will present recent results by Olofsson and Shaw (2001) concerning sampling distributions in the multitype Galton–Watson process. These results allow one to find the expectation and variance of the frequency of particles of a given type in generation n of the multitype Galton–Watson process. These are given in terms of the probability generating function of the offspring distribution. Furthermore, given that a particle of some type is sampled in generation n, the sequence of types of its parent particles in generations $n-1, n-2, \ldots, 2, 1, 0$ is a discrete inhomogeneous Markov chain with different transition probabilities at each step. These results simplify simulations of genealogies and accumulated mutations in at least two interesting biological models (Sections 6.7 and 6.8).

The approach taken in Olofsson and Shaw (2001) is similar to that used in a sequence of articles by Waugh (1981) and Joffe and Waugh (1982, 1985, 1986), who addressed the so-called kin number problem in Galton–Watson populations. They establish exact formulas for the probability distributions of family trees of a randomly sampled individual in a fixed generation. The most extensive treatment is of the single-type case (Joffe and Waugh, 1982); the multitype case is addressed in Joffe and Waugh (1985, 1986).

We will use the notation ψ_n for the probability generating function of $(Z_n^0, \ldots, Z_n^r)$ when there is an arbitrary number of ancestors $(Z_0^0, \ldots, Z_0^r)$, reserving the notation f_n^i for the case of one single ancestor of type i.

6.6.1 Formulas for mean and variance

The following result gives the mean and variance of the proportion of type-i individuals in the nth generation, conditioned on this generation being nonempty. We use the notation $|\mathbf{Z}_n|$ for the total number of individuals in the nth generation (i.e., $|\mathbf{Z}_n| = \sum_{k=0}^{r} Z_n^k$).

Theorem 25. *Let* $\mathbf{u}$ *be a vector with all* u *entries except for a* v *in the* i*th position [*$\mathbf{u} = (u, \ldots v, \ldots, u)$*, and* $\mathbf{0} = (0, 0, \ldots, 0)$*] and denote by* ψ_n *the joint probability generating function of* $(Z_n^1, \ldots, Z_n^r)$*. Then,*

$$\mathrm{E}\left[\frac{Z_n^i}{|\mathbf{Z}_n|}\middle| |\mathbf{Z}_n| > 0\right] = \frac{1}{1-\psi_n(\mathbf{0})}\int_0^1 \frac{\partial}{\partial v}\psi_n(\mathbf{u})\bigg|_{u=v=s} ds$$

and

$$\mathrm{Var}\left[\frac{Z_n^i}{|\mathbf{Z_n}|}\middle| |\mathbf{Z}\right]$$

$$= \frac{1}{1-\psi_n(\mathbf{0})}\int_0^1 -\log s\left(s\frac{\partial^2}{\partial v^2}\psi_n(\mathbf{u})\bigg|_{u=v=s} + \frac{\partial}{\partial v}\psi_n(\mathbf{u})\bigg|_{u=v=s}\right) ds$$

$$-\left(\frac{1}{1-\psi_n(\mathbf{0})}\int_0^1 \frac{\partial}{\partial v}\psi_n(\mathbf{u})\bigg|_{u=v=s} ds\right)^2.$$

The methods of proof are inspired by those of Joffe and Waugh (1985, 1986). Details of the proof are described in Olofsson and Shaw (2001).

6.6.2 The Markov property

Next, we investigate the dependence structure in the sequence of types in the lineage of a particle in the nth generation. We may think of this particle as sampled at random and denote its type by T_n. Because

$$\mathrm{P}(T_n = i) = \mathrm{E}\left[\frac{Z_n^i}{|\mathbf{Z}_n|}\middle| |\mathbf{Z}_n| > 0\right]$$

the probability $P(T_n = i)$ can be obtained from Theorem ??. Denote the type of this particle's parent by T_{n-1}, its grandparent's type by T_{n-2}, and so on; we thus obtain a sequence of types $T_n, T_{n-1}, \ldots, T_0$, the type of the ancestor. It turns out that, conditional on nonextinction, this sequence is a nonhomogeneous Markov chain with transition probabilities given by a formulas invoking the probability generating function of the offspring distribution. This can be utilized for simulations to assess the type variation in lineages of sampled particles. Let φ_{ij} denote the probability generating function of the number of j-type offspring of an i-type parent; that is,

$$\varphi_{ij}(s) = \mathrm{E}_i[s^{X^{(j)}}] = f^i(1, \underbrace{\ldots, s, \ldots}_{j\text{th argument}}, 1).$$

Let ψ_k be as in Theorem 25.

Theorem 26. *The sequence of types $T_n, \ldots, T_0$ in the genealogy of an individual randomly sampled from generation n is a nonhomogeneous Markov chain with transition probabilities*

$$\begin{aligned} &\mathrm{P}(T_k = i | T_{k+1} = j) \\ &= \frac{1}{1 - P\left(Z_{k+1}^{(j)} = 0\right)} \int_0^1 \frac{\partial}{\partial v} \psi_k \left(\varphi_{0j}(u), \ldots, \varphi_{ij}(v), \ldots, \varphi_{rj}(u)\right) \Big|_{u=v=s} ds \end{aligned}$$

where

$$\mathrm{P}\left(Z_{k+1}^{(j)} = 0\right) = \psi_k \left(\varphi_{0_j}(0), \varphi_{1_j}(0), \ldots, \varphi_{r_j}(0)\right).$$

Note that there is a v in the ith position and u in the other positions in the argument of ψ_k.

Details of the proofs of both theorems are described in Olofsson and Shaw (2001). If the branching process is irreducible, the backward Markov chain becomes asymptotically homogeneous in the sense that as $n \to \infty$, the transition probabilities converge to limiting distributions. This follows from the convergence theorem of Jagers (1991), where convergence toward the so-called stable population is investigated. In the PCR application in Section 6.8, this can be observed empirically already for low values of n.

6.7 *Application*: Deletions in Mitochondrial DNA

Mitochondria are organelles in cells carrying their own DNA. Just like nuclear DNA, mitochondrial DNA (mtDNA for short) is subject to mutations which may take the form of base substitutions, duplications, or deletions. This application focuses on one particular mutation, the mtDNA^{4977} deletion. This is a mutation which causes a deletion of about one-third of the mitochondrial genome, thus causing a DNA molecule which is significantly smaller than normal. It has been observed that high levels of deletions are associated with certain degenerative diseases – for example, Kearns–Sayre syndrome (Chinnery and Turnbull 1999). These levels may be as high as 40–50%. Low levels (0.5–12%) have been observed in different regions of the brain of healthy humans. There is a wide variety of issues involved, such as different levels in different types of tissue, but we will not attempt to address any of these. Instead, we focus on how the process of replication of mitochondrial DNA can be described as a multitype Galton–Watson process and how the sampling formulas of the previous sections can be applied to explore how deletions accumulate over time. The idea to use branching processes in this application was first described in Shenkar et al. (1996) and in the unpublished work of Navidi et al. (1996).

The population of mitochondrial DNA is modeled as a two-type process where the types are 0 (normal) and 1 (mutant). A normal can give birth to either two normals or, if there is a mutation, one normal and one mutant. The latter happens with probability λ and we refer to λ as the mutation rate. Mutants can only give birth to mutants. A DNA molecule also may die without reproducing [so-called mitochondrial turnover; see Arking (1998)] and we let the survival probabilities be p and q for normals and mutants, respectively. This gives the following offspring distributions:

$$p_0(0,0) = 1 - p, \qquad p_0(2,0) = p(1-\lambda), \qquad p_0(1,1) = p\lambda$$

for normals and

$$p_1(0,0) = 1 - q, \qquad p_1(0,2) = q$$

for mutants. This gives the joint probability generating functions

$$\varphi_0(u,v) = 1 - p + p\lambda uv + p(1-\lambda)u^2 \tag{6.61}$$

and

$$\varphi_1(u,v) = 1 - q + qv^2. \tag{6.62}$$

The proportion of mutants in the nth generation is

$$\frac{Z_n^{(0)}}{Z_n^{(0)} + Z_n^{(1)}}$$

and we can use Theorem 25 to compute its expectation and variance. Further details are described in Olofsson and Shaw (2001).

6.8 *Application*: Polymerase Chain Reaction

This application can be understood as a sequel to Section 1.2. As described in that section, a DNA molecule in any given cycle of PCR either existed before the cycle or is newly created (this is the essence of the semiconservative replication). The process is modeled as a two-type process where the type space is $\{0, 1\}$, 0 for "old" and 1 for "new." The distinction is crucial to mutation studies since because mutations only arise on newly created particles. The offspring distribution is

$$p_0(1,0) = 1 - p, \qquad p_0(1,1) = p,$$
$$p_1(1,0) = 1 - p, \qquad p_1(1,1) = p,$$

where p is the cycle efficiency (i.e., the probability that a given molecule replicates successfully in a given PCR cycle). This leads to joint probability generating functions

$$\varphi_0(u, v) = \varphi_1(u, v) = (1 - p)u + puv.$$

For the simulations, Theorem 25 was used to compute the distribution of a randomly sampled particle in generation n, and Theorem 26 was used to compute the transition probabilities. Simulations were then performed in which a particle was sampled at random from generation n and the sequence of types in its lineage back to the ancestor generated. Each time a particle of type 1 appeared, it was independently assigned a new mutation with probability λ. The values $n = 30$, $p = 0.7$, and $\lambda = 0.05$ were used [see Section 1.2, where, however a slightly different notation, consistent with the Weiss and von Haeseler (1997) article, was used].

Olofsson and Shaw (2001) show a histogram of the number of mutations in the lineage of a randomly sampled particle in generation 30, based on 100,000 simulation runs of the Markov chain. The transition probabilities $P(T_k = i | T_{k+1} = j)$ converge to a limiting distribution as $n \to \infty$, and in this particular application, the convergence is rapid. The limiting transition probabilities can be computed as

$$P(T_k = i | T_{k+1} = j) = \frac{\nu(i)M(i, j)}{\rho\nu(j)},$$

where $M(i, j) = E_i[X^{(j)}]$, the (i, j)th entry in the mean reproduction matrix

$$M = \begin{pmatrix} M(0,0) & M(0,1) \\ M(1,0) & M(1,1) \end{pmatrix} = \begin{pmatrix} 1 & p \\ 1 & p \end{pmatrix},$$

ρ is the largest eigenvalue of M, and ν is the left eigenvector of M corresponding to ρ. In this case,

$$\rho = 1 + p, \qquad \nu(0) = \frac{p}{1+p}, \qquad \nu(1) = \frac{1}{1+p},$$

which gives, in the limit,

$$\mathrm{P}(T_k = 1 | T_{k+1} = j) = \frac{p}{1+p} \approx 0.41$$

for both $j = 0$ and $j = 1$. The computations reveal that this limit is reached after less than 10 generations. Still further details may be found in the unpublished dissertation by Shaw (2000).

6.9 Other Works and Applications

6.9.1 Hemopoiesis and clonal cell populations

Multitype branching processes are the natural tool to model the proliferation of cells undergoing differentiation (i.e., changing gene expression, morphology, and biological function). Usually, as it is the case in the hemopoietic (blood-production) system, populations of differentiating cells are organized in nets. The top population of the net is formed by stem cells, which can either produce progeny of its own type or of any other type. Each next population is committed to differentiation in some direction (i.e., it can produce progeny of its own type or of a limited subset of types, usually just one type of more mature cells). The bottom population(s) are not capable of proliferation. This type of multitype branching process is called reducible. Early articles employing branching-type models are Till et al. (1964) and Vogel et al. (1969). A simulation model was developed by Rittgen (1983).

The stochastic model of mast cells proliferation developed by Pharr et al. (1985) assumed a two-type Galton–Watson process including proliferative and nonproliferative cells. Each proliferative cell gives rise to either two proliferative progeny, or to two nonproliferative progeny, or it may die. Each nonproliferative cell may either survive (i.e., produce one progeny) or die. In principle, this model is identical to that of Section 3.2. Predictions of the model by Pharr et al. (1985) were fitted to colony size data, with a good agreement. In a further article (Nedelman et al. 1987), the model was extended to a Bellman–Harris process and maximum likelihood was used to estimate parameters.

A more general model including a chain of maturing cell populations, of the type described above, was designed by Ciampi et al. (1986) to model proliferation of human ovarian carcinoma cells. Modeling, using a multitype Galton–Watson process, involved calculating the asymptotic distributions of colony sizes and data-based estimation of the self-renewal probability of stem cells. This latter is the conditional probability of a stem cell producing two stem cells (as opposed to producing two differentiated cells), given that it does not die or rest. The self-renewal probability is a parameter of potential diagnostic value (also, see Loeffler and Wichmann 1980). Another related reference is the book by Macken and Perelson (1988), which considers multitype Galton–Watson models of the hemopoietic (blood-production) system although without much reference to data. Therneau et al. (1989) modeled early stages of development of cell colonies using symbolic calculations to iterate the probability generating functions of the process.

Articles by Stivers and Kimmel (1996a, 1996b) and Stivers et al. (1996) concerned the observed inheritance of sizes of primary and secondary colonies in experiments by Axelrod et al. (1993) and Gusev and Axelrod (1995), discussed in Section 5.5.1. The main question is to determine what modes of inheritance of cell lifetimes are consistent with the observed correlations of the sizes of primary and secondary colonies, which are positive, equal to approximately 0.6, and, at the same time, consistent with the observed variances in colony sizes. Various modes are considered, including "clonal," in which the life-length distribution of the secondary colony is affected by the lifetime of its founder drawn at random from a primary colony (Stivers and Kimmel 1996a, 1996b). Another variant is generational inheritance in a model in which two types of cells, with differing proliferative potential, can differentiate into each other (Stivers et al. 1996). This latter model provides a fit to the observations.

Abkowitz et al. (1996) used experimental data and branching process simulations to demonstrate that hemopoiesis, the process of blood cell production, has a random nature. They use results of irradiation experiments carried out on Safari cats, a race being a cross between domestic cats and wild Geoffroy cats. These two species of cats have evolved independently and have electrophoretically distinct phenotypes of the X-chromosome-linked enzyme glucose-6-phosphate dehydrogenase (G6PDH). Female Safari cats are generally balanced heterozygotes with, on average, equal numbers of progenitor and differentiated blood cells of each parental phenotype. However, females deprived of their bone marrow by irradiation and then given autologous transplants of 30 quiescent hematopoietic cells end up, after a period of fluctuations, with variable proportions of progenitors of each parental phenotype. The pattern of variability is consistent with simulations based on a multitype branching process model. This article, although it contains no mathematics, provides important arguments concerning the applicability of branching processes.

6.9.2 *Gene amplification*

The biological introduction to gene amplification can be found in Section 2.1.6. Mathematical models are described in Sections 3.6, 6.1, 7.1, and 7.4. Other authors considered diverse aspects of gene amplification (Seneta and Tavaré 1983). Harnevo and Agur (1991) constructed a comprehensive model of gene amplification in the form of a multitype branching process, in the context of resistance of cancer cells to cytotoxic drugs. The number of types is finite, as these authors assume limits to the number of copies of the amplified gene. Theoretical considerations are followed by a modeling study in which the dynamics of growth of cells with amplified phenotype (drug resistant) is followed. A similar mathematical model was employed by Harnevo and Agur (1992) to explore various strategies of cancer chemotherapy, assuming that the main mechanism of drug resistance was gene amplification. Harnevo and Agur (1993) contains a critical discussion of approaches to modeling of the gene amplification process, including the model of Kimmel et al. (1992), described in Section 7.1. Kimmel (1994) is a review.

Finally, we should mention the article by Peterson (1984). This is an article in which evidence is collected suggesting that expression of many proteins in cells occurs at levels which form an arithmetic or geometric progression. Peterson (1984) postulated that this may be due to the variable number of copies of respective genes, present in a given cell (Quantitative Shift Model). Although this article does not contain mathematics, a multitype branching process is implicitly involved.

6.9.3 Modeling in varying environments

Modeling of multitype branching processes in varying environments is important when we consider populations of biological cells subject to external controls. The usual backward (retrospective) technique of decomposition into subprocesses generated by first-generation progeny leads to pgf and moment equations which tie together processes started at different times and therefore different, so that self-recurrence cannot be used. This is inconvenient if the solution is to be extended step-by-step, (e.g., by numerical procedures), as it is the case in modeling of cancer chemotherapy. For this reason, it is desirable to develop a forward (prospective) technique, which would provide a recurrence or equation allowing to continue the pgf or moment characterization of the process in time. In Kimmel (1982), it is demonstrated that an equivalent (dual) set of integral equations exists, which allows prospective continuation of the expectations of the process in time. In Kimmel (1983), it is shown that a prospective equation of a kind can be written not for the probability generating functions, but for the probability generating functional which describe the multivariate point process of births and deaths in the branching process.

Klein and Macdonal (1980) consider a multitype Markov process in periodic environment.

CHAPTER 7

Branching Processes with Infinitely Many Types

In this chapter, we consider a number of examples of branching processes with infinite-type spaces. No systematic theory can be presented. However, in Section 7.6, we review various approaches generalizing the denumerable case. Also, general processes (Appendix C.1) include the denumerable-type space as a special case. We will base considerations on an analogy with the finite mutitype case whenever possible. However, the stress is on interesting and diverse properties, which are different from the finite multitype setup, and on biologically motivated examples.

We will begin by presenting an example of a stable process, using another variant of the gene amplification model (Section 7.1). The subsequent example is the reducible process of loss of telomere endings, which displays polynomial dynamics (Section 7.2). Sections 7.3–7.5 deal with the problem of quasistationarity in the context of branching random walks and branching-within-branching. These examples can be understood as generalizations of Yaglom's Theorem (Theorem 7). Finally, in Section 7.7, we develop a series of structured population models, which can be classified as branching processes with continuous-type spaces.

7.1 *Application*: Stable Gene Amplification

This is a model for a variant of the process of gene amplification different from the one considered in the chapter on the Galton–Watson process. The previous model accounted for the instability of some amplified genes by their loss from cells during cell division. The loss of these extrachromosomal elements was associated with the lack of centromeres which are found on chromosomes and are required for faithful segregation at cell division.

The model developed here accounts for situations in which gene amplification can be either stable or unstable. It is based on different experimental observations and a more extensive biological model (Windle et al., 1991). It describes the initiation of amplification as the breakage of a piece of a chromosome releasing a fragment containing a gene or genes but not a centromere. Genes on these acentric extrachromosomal fragments may replicate and recombine, forming increased numbers of tandemly repeated genes. As in the previous model, the extrachromosomal elements may segregate, although their segregation is not faithful because they do not have the required centromeres.

A new aspect of the model is that it also includes the possibility of stabilization of the number of amplified genes following their reintegration into chromosomes. This is because reintegration links the amplified genes to the centromeres on chromosomes, allowing them to be faithfully segregated at cell division.

From the mathematical viewpoint, this model is an example of a decomposable process. Decomposable processes include a subclass of transient types which are irreversibly lost from the process. Processes limited to the remaining types may behave variously. In this particular application, as we will see, it will reach a limit distribution. A decomposable process cannot be positive regular, as a whole, although the persistent subprocess may be.

7.1.1 Assumptions

The following is the list of model hypotheses (Fig. 7.1):

1. All acentric (extrachromosomal) elements evolve independently of each other.
2. Types:
   - **a.** Acentric elements containing $i = 1, 2, \ldots$ gene copies
   - **b.** Chromosomes with one or more sites containing reintegrated elements, each containing $i = 1, 2, \ldots$ gene copies.
3. In each cell generation, three types of events can occur for each acentric extrachromosomal element:
   - **a.** Element replicates and breaks at a random site, and the pieces segregate
   - **b.** Element replicates and does not break
   - **c.** Element reintegrates into a chromosome.
4. With probability a, the element with i gene copies replicates and yields a single element with $2i$ gene copies and then breaks at a random site producing two pieces with lengths j and $2i - j$, where $j = 1, \ldots, 2i - 1$. The probability of breakage at each site is the same and equal to $1/(2i - 1)$. The pieces segregate so that they both pass to the same progeny cell with probability α and pass to different progeny cells with probability $1 - \alpha$.
5. With probability b, the element with i gene copies replicates to yield a single element with $2i$ gene copies, but it does not break. It then passes with probability $\frac{1}{2}$ to one of the two progeny.
6. With probability c, the element containing i copies of the gene, is integrated into a chromosome with a centromere and then replicates and segregates with

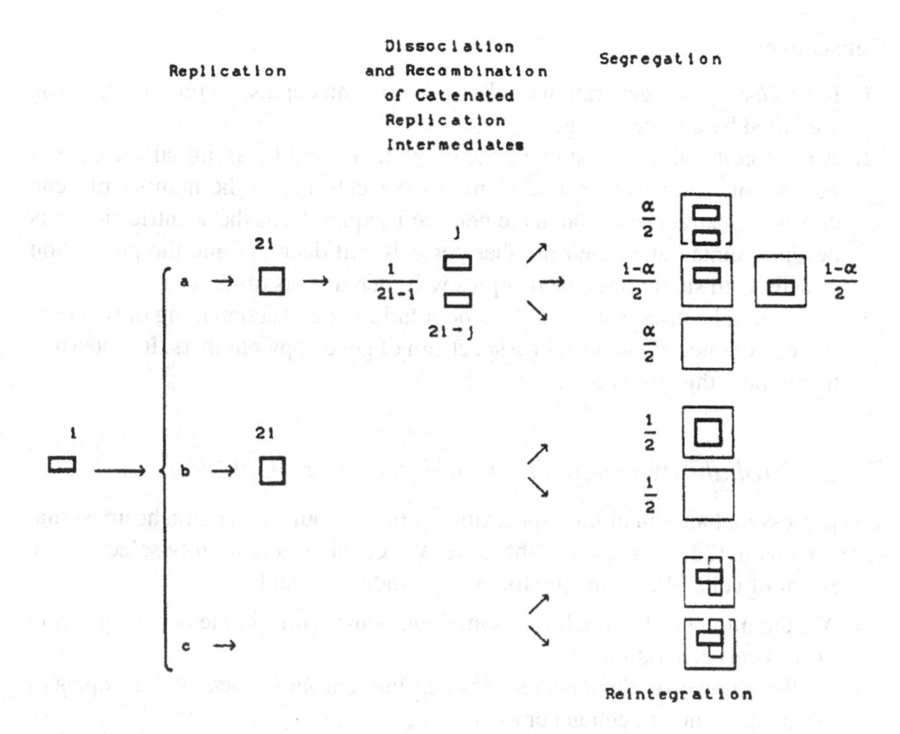

FIGURE 7.1. Schematic representation of the events in the gene amplification model. Source: Kimmel, M., Axelrod, D.E. and Wahl, G.M. 1992. A branching process model of gene amplification following chromosome breakage. Mutation Research 276: 225–239. Figure 1, page 229. Copyright: 1992 Elsevier Science Publishers B.V.

the chromosome. This results in progeny cells with equal number of gene copies. No further breakage or increase or decrease in gene copy number occurs at this site at subsequent cell divisions. The probability of reintegration is equal to $c = 1 - (a + b)$.

Initial conditions

At the beginning of the process, a single cell contains a single extrachromosomal element with one gene copy, $i = 1$. It is understood that this element was formed in the past by deletion of one copy of a chromosomal gene in a founder cell.

Remarks

1. Breakage can be understood as imperfect resolution of replicated DNA.
2. If breakage occurs, a randomly chosen progeny cell will contain both pieces with probability $\alpha/2$, no piece with probability $\alpha/2$, a single piece of size j with probability $(1 - \alpha)/2$, and a single piece of size $2i - j$ with probability $(1 - \alpha)/2$.

Consequences

1. In successive cell generations, cells with no gene copies are produced. These are killed by a selective agent.
2. Among cells with at least one gene copy, there will be an initial increase in number of extrachromosomal elements per cell and in the number of gene copies per extrachromosomal element. Subsequently, as the acentric elements become reintegrated, their number per cell will decrease and the proportion of cells with stably integrated copies will increase (as observed).
3. An eventual consequence will be a population of cells containing only one or more integrated elements with a spectrum of gene copy numbers. It is possible to compute this distribution.

7.1.2 Probability generating functions and expectations

The process includes an infinite spectrum of chromosomal and extrachromosomal elements with $1, 2, \ldots$ copies of the gene. We consider a randomly selected line of descent of cells. We define the following random variables:

- X_n^i, the number of extrachromosomal elements with i copies of the gene, in the nth cell generation
- Y_n^i, the number of elements reintegrated into chromosomes, with i copies of the gene, in the nth cell generation

The sequence $\{\{(X_n^1, Y_n^1), (X_n^2, Y_n^2), \ldots\}, n = 0, 1, 2, \ldots\}$ is a multitype Galton–Watson process with a denumerable infinite number of particle types.

Let us consider the aggregated process $\{(X_n, Y_n),\ n = 0, 1, 2, \ldots\}$, where

$$X_n = \sum_{i=1}^{\infty} X_n^i, \qquad Y_n = \sum_{i=1}^{\infty} Y_n^i$$

are the total number of the extrachromosomal elements and elements reintegrated into chromosomes, in generation n. Following the rules of the process (see Fig. 7.1), the pgf of the number of progeny of an extrachromosomal element is equal to

$$f^1(s_1, s_2) = \frac{a\alpha}{2} s_1^2 + \left[a(1-\alpha) + \frac{b}{2}\right] s_1 + \frac{a\alpha + b}{2} + cs_2. \tag{7.1}$$

The coefficient of the quadratic term reflects the fact that two extrachromosomal elements are produced from a single one only if breakage occurs (wp a), both elements segregate into one progeny (wp α), and this progeny belongs to the lineage followed (wp $\frac{1}{2}$). The remaining coefficients are derived analogously.

The pgf of the number of progeny of a reintegrated element is simply $f^2(s_2) = s_2$, as such element is stable. Let us denote by $f_n^1(s_1, s_2)$ the joint pgf of $\{(X_n, Y_n)|(X_0, Y_0) = (1, 0)\}$. The following relationship can be derived using the backward approach as in Section 3.2:

$$f_{n+1}^1(s_1, s_2) = \frac{a\alpha}{2}[f_n^1(s_1, s_2)]^2 + \left[a(1-\alpha) + \frac{b}{2}\right] f_n^1(s_1, s_2) + \frac{a\alpha + b}{2} + cs_2. \tag{7.2}$$

This equation provides a recursive procedure for finding distributions of the process.

The process tends to a nontrivial limit with probability 1,

$$(X_n, Y_n) \longrightarrow (0, Y_\infty) \quad \text{wp } 1, \text{ as } n \to \infty. \tag{7.3}$$

To demonstrate it, let us first note that $Y_{n+1}(\omega) \geq Y_n(\omega)$, which yields the almost sure convergence of Y_n, with the limit being possibly an improper random variable. However, passing to infinity with n in Eq. (7.2) yields the following quadratic equation for the pgf of Y_∞:

$$\frac{a\alpha}{2}[f^1_\infty(1, s_2)]^2 + \left(a(1-\alpha) + \frac{b}{2} - 1\right) f^1_\infty(1, s_2) + \frac{a\alpha + b}{2} + cs_2 = 0. \tag{7.4}$$

The pgf solution of Eq. (7.4) verifies $f^1_\infty(1, s_2)|_{s_2=1} = 1$, which means that Y_∞ is a proper random variable. On the other hand, if we set $s_2 = 1$ in Eq. (7.1), we see that $\{X_n\}$ is a subcritical Galton–Watson process, which yields $\mathrm{P}\{\lim_{n\to\infty} X_n = 0\} = 1$. This completes the proof of the property in expression (7.3).

Let us note that in an experimental setting, only cells with a nonzero number of elements (extrachromosomal or reintegrated) are observed, as only these cells survive under drug selection pressure. Therefore, all distributions should be conditional on nonextinction of the process [i.e., on the event $\{(X_n, Y_n) \neq (0, 0)\}$]. In particular, the conditional probability that, in generation n, extrachromosomal elements are still present in the lineage is

$$r_n = \mathrm{P}\{X_n > 0 | (X_n, Y_n) \neq (0, 0)\} = \frac{1 - f^1_n(0, 1)}{1 - f^1_n(0, 0)}. \tag{7.5}$$

Let us consider the expectations of the complete process $\{\{(X^1_n, Y^1_n), (X^2_n, Y^2_n), \ldots\}, n = 0, 1, 2, \ldots\}$,

$$\mu^i_n = \mathrm{E}(X^i_n), \qquad \nu^i_n = \mathrm{E}(Y^i_n), \quad i \geq 1,\ n \geq 0. \tag{7.6}$$

It is not difficult to verify that the infinite vector $\mu_n = \{\mu^1_n,\ \mu^2_n, \ldots\}$ satisfies the recurrence

$$\mu_{n+1} = \mu_n \mathcal{M}, \quad n \geq 0, \qquad \mu^i_0 = \delta_{1i}, \tag{7.7}$$

where $\mathcal{M}$ is an infinite matrix of the form

$$\mathcal{M} = \begin{pmatrix} a & b/2 & 0 & & & & & & & \mathbf{0} \\ a/3 & a/3 & a/3 & b/2 & 0 & & & & & \\ a/5 & a/5 & a/5 & a/5 & a/5 & b/2 & 0 & & & \\ a/7 & a/7 & a/7 & a/7 & a/7 & a/7 & a/7 & b/2 & 0 & \vdots \\ \vdots & \vdots & \vdots & \vdots & \vdots & \vdots & \vdots & \vdots & \vdots & \vdots \end{pmatrix}. \tag{7.8}$$

The expectations of Y^i_n satisfy

$$\nu^i_{n+1} = \nu^i_n + c\mu^i_n, \quad n \geq 0,\ \nu^i_0 = 0. \tag{7.9}$$

Let us note that $\nu_{n+1}^i \geq \nu_n^i$. Using this and an analysis involving Eq. (7.4), we obtain $\lim_{n\to\infty} \sum_{i=1}^{\infty} \nu_n^i = \lim_{n\to\infty} \mathrm{E}(Y_n) < \infty$. Therefore, $\lim_{n\to\infty} \nu_n^i$ exists and is finite ($i \geq 1$).

The expectations $\mu_n^i = \mathrm{E}(X_n^i)$ and $\nu_n^i = \mathrm{E}(Y_n^i)$, properly normed, can be understood as distributions of the sizes of extrachromosomal and reintegrated elements in the lineage. To calculate these expectations, it is convenient to introduce the generating functions

$$M_n(z) = \sum_{i=1}^{\infty} \mu_n^i z^i, \qquad N_n(z) = \sum_{i=1}^{\infty} \nu_n^i z^i, \quad z \in [0, 1].$$

Equations (7.7) and (7.9) yield the following relationships for the generating function

$$M_{n+1}(z) = \frac{b}{2} M_n(z^2) + \frac{az}{1-z} \int_z^1 \frac{M_n(u^2)}{u^2}\, du, \ n \geq 0,$$

where $M_0(z) = z$, and

$$N_n(z) = c \sum_{k=0}^{n-1} M_k(z).$$

After carrying out differentiations with respect to z in the first of the above relationships, we obtain that the mean size of the extrachromosomal elements in generation n is equal to

$$\frac{M_n'(1-)}{M_n(1)} = \left(\frac{b+a}{\frac{b}{2}+a} \right)^n,$$

which tends to ∞ as $n \to \infty$. The expected size of reintegrated elements has a finite limit

$$\frac{N_\infty'(1-)}{N_\infty(1)} = \frac{1-a-b/2}{1-a-b}.$$

7.1.3 *Model versus data*

Parameters of the model for a single experimental system can be deduced based on experiments by Geoffrey Wahl and colleagues of the Salk Institute (Windle et al. 1991). From their results, it is possible to estimate the following quantities:

1. The fraction of nonextinct cells still containing extrachromosomal elements after nine generations ($r_9 \sim 0.39$)
2. The fraction of nonextinct cells still containing extrachromosomal elements after 35 generations ($r_{35} \sim 0.02$, highly inaccurate)
3. The fraction of nonextinct cells with one or two elements extrachromosomal and/or reintegrated (as opposed to cells containing $\geq$ three elements), after nine generations ($p_{12} \sim 0.54$).

The theoretical relationships (7.5) and (7.2), with parameter values

$$\alpha = 1, \quad a = 0.92, \quad b = 0.035,$$

yield

$$r_9 = 0.39, \quad r_{35} = 0.05, \quad p_{12} = 0.63,$$

in approximate agreement with the experiment.

The expected size of reintegrated elements predicted by the model is rather low, equal to $\frac{4}{3}$. However, this is based on the assumption that the initial extrachromosomal element in generation 0 is of a unit size.

7.2 *Application*: Mathematical Modeling of the Loss of Telomere Sequences

7.2.1 *Stochastic model*

Telomeres are structures at the ends of chromosomes. They consist of repeated DNA sequences which play a role in replication of the ends of DNA and in preventing the ends of chromosomes from sticking together. The number of repeat sequences of the telomeres is variable and, on the average, declines with the increasing number of divisions of normal cells in culture and of somatic cells in organisms (Larson et al. 1987). Reviews on the biology of telomeres include Blackburn (1991), Greider and Blackburn (1996), Greider (1996), Harley (1991), and Zakian (1995, 1996).

Cellular senescence, the loss of capacity to proliferate, seems to be associated with the inability to maintain a minimum number of telomere sequences. In contrast, immortalized cells such as cancer cells, seem to be able to maintain a low but effective number of telomere sequences. The first researcher who noted the relationship between telomere endings and cell senescence was Olovnikov (1973). He correctly attributed this loss to the end-replication problem, which arises because of the inability of the DNA polymerase to replicate the downstream end of the DNA molecule. The effect is that each successive DNA replication results in a copy, which is shorter at one end.

Our model (Arino et al., 1995) describes shortening of telomeres by incomplete replication. The two uses of the model are predictions of (1) the expected telomere length and (2) of the fraction of viable cells in aging cell populations. For these purposes, it is first necessary to describe the dynamics of telomere loss from a single chromosome. For simplicity, we proceed as if the process of telomere loss ended when all the telomere deletion units, each containing possibly more than a single DNA repeat, are lost. The same mathematics applies to telomere loss until a particular checkpoint is encountered.

A chromosome is an entity with a centromere, whereas a chromatid is a double helix composed of two single strands of DNA. In the G_1 phase of the cell cycle, before DNA replication, a chromosome is composed of one chromatid, whereas

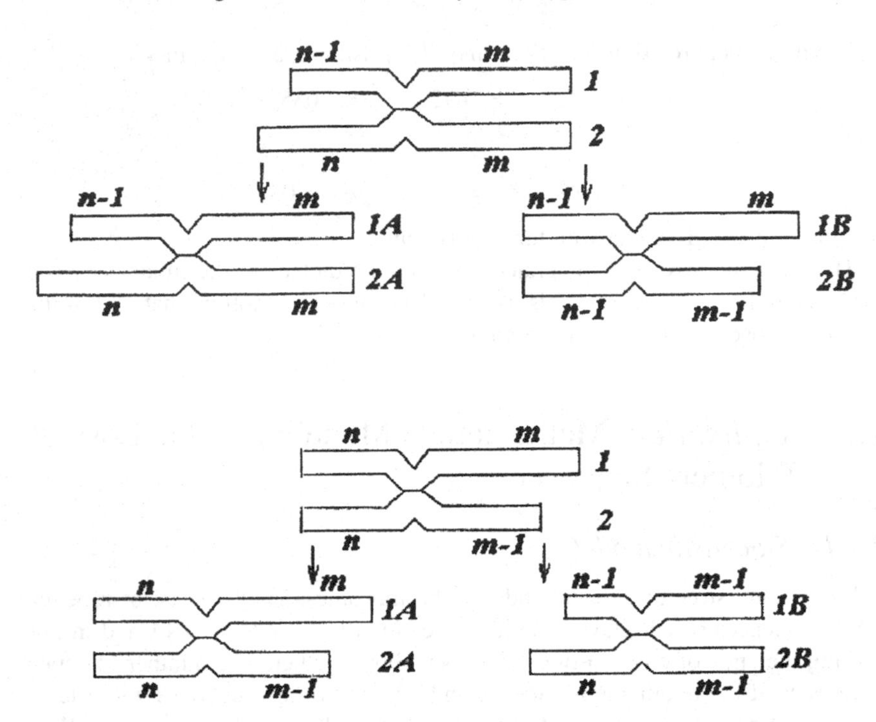

FIGURE 7.2. Transition rules for deletion and segregation of telomere ends on a chromosome in G_1. DNA strands 1 and 2 replicate and segregate into different daughter cells A and B, resulting in chromatids (1A, 2A) and (1B, 2B), respectively. Due to the end-replication problem, one DNA strand on each of the newly created chromatids contains additional deletion at its right or left end, depending on its orientation and presence of the deletion on the corresponding strand of the mother chromatid. For additional explanations, compare Fig. 1 in Levy et al. (1992). Source: Arino, O., Kimmel, M., and Webb, G.F. 1995. Mathematical modeling of the loss of telomere sequences. Journal of Theoretical Biology 177: 45–57. Figure 1, page 46. Copyright: 1995 Academic Press Limited.

in the G_2 and M phases, after DNA replication, a chromosome is composed of two chromatids. Levy et al. (1992) described telomere loss in terms of what happens to single DNA strands in G_1. We follow that description. Figure 7.2 depicts the scheme of deletion and segregation of telomere sequences on chromosome ends. It can be summarized mathematically as follows:

- Each chromatid is composed of two strands named upper or $5' \to 3'$, and lower or $3' \to 5'$, each of which has two ends named left and right. The numbers of telomeric deletion units on both ends of both strands are symbolically represented by quadruples of the form $(a, b; c, d)$, where a and c correspond to the left and right ends of the upper strand, respectively, and b and d correspond to the left and right ends of the lower strand, respectively. The only important combinations of a, b, c, and d are of the form $(n-1, n; m, m)$ or $(n, n; m, m-1)$ because they always arise after a single replication round (details not shown).

- Cells containing chromatids described by the quadruple $(n-1, n; m, m)$ give birth to two progeny containing chromatids of the types $(n-1, n; m, m)$ and $(n-1, n-1; m, m-1)$, respectively. This transition rule as well as the dual rule for the other admissible type are depicted symbolically below. Let us note that one progeny is always of the same type as the parent cell, whereas the other is missing two sequences, each on a different end of a different strand:

$$(n-1, n; m, m) \begin{cases} \rightarrow & (n-1, n; m, m) \\ \rightarrow & (n-1, n-1; m, m-1), \end{cases}$$
$$(n, n; m, m-1) \begin{cases} \rightarrow & (n, n; m, m-1) \\ \rightarrow & (n-1, n; m-1, m-1). \end{cases}$$

- Proliferation ends when the telomere ends become short enough. Without a loss of generality, we assume that cells of the types $(n-1, n; 0, 0)$ and $(0, 0; m, m-1)$ have a single progeny of the type identical to that of the parent cell; that is,

$$\begin{aligned} (n-1, n; 0, 0) &\rightarrow (n-1, n; 0, 0), \\ (0, 0; m, m-1) &\rightarrow (0, 0; m, m-1). \end{aligned}$$

If we renumber states in such way that index $k = 0, 1, \ldots$ is equal to the sum of numbers of deletion units on the left ends of the upper and lower strand and index $l = 0, 1, \ldots$ is equal to the sum of numbers of deletion units on the right ends of the upper and lower strand:

$$k = \begin{cases} 2n & \text{if } (n, n; m, m-1) \text{ occurs} \\ \text{or} & \\ 2n-1 & \text{if } (n-1, n; m, m) \text{ occurs,} \end{cases} \tag{7.10}$$

$$l = \begin{cases} 2m & \text{if } (n-1, n; m, m) \text{ occurs} \\ \text{or} & \\ 2m-1 & \text{if } (n, n; m, m-1) \text{ occurs,} \end{cases} \tag{7.11}$$

then the admissible transitions become

$$(k, l) \begin{cases} \rightarrow & (k, l) \\ \rightarrow & (k-1, l-1), \end{cases} \tag{7.12}$$

$$(k, 0) \rightarrow (k, 0), \tag{7.13}$$

$$(0, l) \rightarrow (0, l). \tag{7.14}$$

In the array (k, l), where k and l are non-negative integers, the admissible transitions belong to disjoint paths which can be numbered by $k - l$ (path number assuming values from $-\infty$ through ∞). Each of these paths can be treated separately. The state number within each path can be taken as $i = \min(k, l)$. Biologically, it is the number of deletion units on the shorter, and therefore limiting, end. Now the transitions have the form

$$i \begin{cases} \to & i \\ \to & i-1, \end{cases} \tag{7.15}$$

$$0 \quad \to \quad 0. \tag{7.16}$$

7.2.2 Branching process

Let us assume that life lengths of cells are independent identically distributed random variables with distribution with density $g(t)$ and cumulative distribution $G(t)$. Let us denote by

$$X_{ij}(t), \ t \in [0, \infty), \ i, j = 0, 1, \ldots,$$

the family of random variables equal to the number of cells in state j at time t, in the process started at time 0 by a single cell in state i.

Our process can be described as a branching random walk. In our case, this means that the type of the progeny object (chromosome) is either identical with the parental type or it is shortened by a single unit. We have

$$X_{ij}(t) = \begin{cases} X_{ij}(t-\tau) + X_{i-1,j}(t-\tau), & \tau \le t \\ \delta_{ij}, & \tau > t \end{cases} \tag{7.17}$$

for all $j = 0, 1, \ldots, i = 1, 2, \ldots$ and $t \in [0, \infty)$. The above equation expresses the fact that the number of cells in state j at time t, in a process started at time 0 by a single cell in state i, is either equal to δ_{ij} if the ancestor cell is still alive or it is equal to the sum of the numbers of cells in state j at time t in two subprocesses started at time τ (i.e., at the moment of the ancestor's death) by the two progeny of the ancestor, one of which is in state i and the other in state $i-1$. Another equation,

$$X_{0j}(t) = \begin{cases} X_{0j}(t-\tau), & \tau \le t \\ \delta_{0j}, & \tau > t \end{cases} \tag{7.18}$$

for all $j = 0, 1, \ldots$ and $t \in [0, \infty)$ expresses the fact that cells in state 0 do not proliferate.

Let

$$M_{ij}(t) = \mathrm{E}[X_{ij}(t)] \tag{7.19}$$

denote the expected count of cells in state j at time t in a process started by an ancestor of type i. We obtain the following equation for the matrix $M(t) = [M_{ij}(t)]$:

$$M(t) = Ag(t) * M(t) + \bar{G}(t)I, \tag{7.20}$$

where $\bar{G}(t) = 1 - G(t)$ and the symbol "$*$" denotes convolution of matrix functions on $[0, \infty)$, $g(t) * M(t) = \int_0^t g(t-\tau)M(\tau)\, d\tau$, I is the infinite identity matrix, and the infinite matrix, A has the form

$$A = \begin{pmatrix} 1 & 0 & 0 & 0 & \cdots \\ 1 & 1 & 0 & 0 & \cdots \\ 0 & 1 & 1 & 0 & \cdots \\ 0 & 0 & 1 & 1 & \cdots \\ \vdots & \vdots & & \ddots & \ddots \end{pmatrix}.$$

The solution of this *backward* equation can be represented as an infinite series

$$M = \left[\sum_{k \geq 0} (Ag)^{*k} \right] * \bar{G}I = \bar{G}I * \left[\sum_{k \geq 0} (Ag)^{*k} \right].$$

The second series is the solution of a dual *forward* equation

$$M(t) = M(t) * Ag(t) + \bar{G}(t)I, \tag{7.21}$$

which is equivalent to the system

$$M_{ij}(t) = g(t) * [M_{ij}(t) + M_{i,j+1}(t)] + \delta_{ij}\bar{G}(t), \; j = 0, 1, \ldots, \; t \geq 0, \tag{7.22}$$

which can be examined separately for each ancestor's state i. This would not be possible with the backward system.

7.2.3 Analysis in the Markov case

If the cell life length distributions are exponential [i.e., the density has the form $g(t) = \alpha \exp(-\alpha t)$], the system of convolution equations (7.22) is equivalent to the following infinite system of differential equations:

$$\dot{M}_{ij}(t) = \alpha M_{i,j+1}(t), \qquad M_{ij}(0) = \delta_{ij}, \quad j = 0, 1, \ldots, \; t \geq 0. \tag{7.23}$$

This system has an explicit solution

$$M_{ij}(t) = \begin{cases} \dfrac{\alpha^{i-j} t^{i-j}}{(i-j)!}, & 0 \leq j \leq i \\ 0, & j > i. \end{cases} \tag{7.24}$$

Let $M_j(t)$ denote the expected number of cells in state j at time t if the initial expected counts of cells in states $0, 1, \ldots$ were $M_0(0), M_1(0), \ldots$. Expressions for $M_j(t)$ are obtained by combining solutions of Eq. (7.23).

If the initial cells belong to finitely many different states, so that

$$M_j(0) = 0, \; j > N, \tag{7.25}$$

then

$$M_j(t) = \sum_{k=j}^{N} M_k(0) \frac{\alpha^{k-j} t^{k-j}}{(k-j)!}. \tag{7.26}$$

We may note that the polynomial dynamics of the expected values is a consequence of the one-way means of communication between types in the process. This is known as reducibility of the process. Biologically, it is a consequence of the fact that loss of telomere repeats is irreversible.

7.2.4 *Model versus data*

The Markov branching process model was employed to reproduce experimental data on telomere loss. Let us suppose that the number of telomeric repeats in a given chromosome at the time the clonal population growth is initiated ($t = 0$) exceeded the critical (checkpoint) length by d deletion units; that is,

$$M_j(0) = \delta_{jd} N_0. \tag{7.27}$$

As Levy and co-workers (1992) pointed out, it is likely that telomeres on different chromosomes differ in their initial number of TTAGGG repeats. Because only the chromosomes with the shortest telomeres are relevant to replicative senescence, only the deletions on the shorter of these chromosomes' 2 telomeres need to be considered. Suppose that there are k such chromosomes with the same critical d and that they segregate independently and that only one critically short telomere is sufficient to signal the cell cycle exit.

We identified two sources of data useful for modeling. One is the article by Harley and Goldstein (1980) in which fractions $F(d, t)$ of proliferating cells were measured at different times after a clonal culture had been established. These data have been used for modeling by Levy and co-workers (1982) (see their Fig. 6).

Another source is the article by Counter and co-workers (1992) which includes experimental data on the expected telomere lengths (mean number of excess deletion units) $n(t)$.

Our expressions for the expected frequencies of telomere repeat counts on a single chromosome can be combined to yield expressions for $F(d, t)$ and $n(t)$ (details in Arino et al., 1995).

Figure 7.3 depicts the results of modeling of the fraction of viable cells as a function of the number of cell doublings of a clonal culture. Experimental data for two independent cultures of a human fibroblast strain (Harley and Goldstein, 1980; after Levy et al., 1992, Figure 6, modified) are compared to predictions using the Markov branching process model.

The fit has been obtained with parameters $d = 65$ and $k = 40$. Note that the number of chromosomes has to be set equal to $k = 40$ to achieve an acceptable fit, otherwise the decrease in $F(d, t)$ is not sharp enough. This number is not very different from the number of human chromosomes (equal to $2 \times 23 = 46$), which may be taken to mean that all chromosomes have the same critical d-value.

Figure 7.4 depicts the results of modeling the mean length of terminal restriction fragments (TRFs) in the function of the number of cell doublings. Experimental data for a number of cultures of normal and transfected cells, up to the crisis time (from Counter et al., 1992) are compared to predictions using the Markov

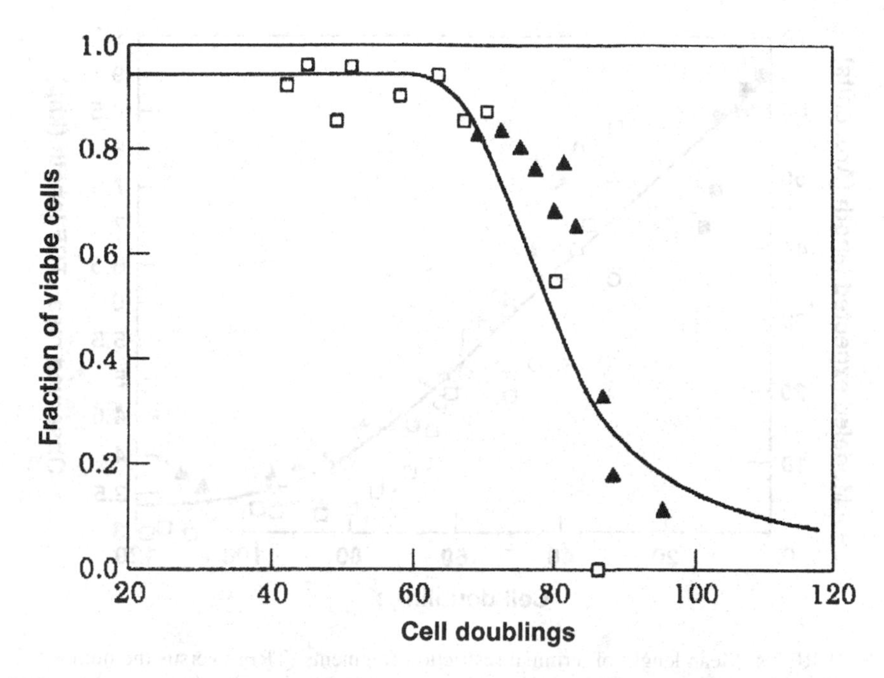

FIGURE 7.3. Fraction of viable cells versus the number of cell doublings. Experimental data for two independent cultures of a human fibroblast strain represented by triangles and squares, are compared to predictions using the Markov branching process model (continuous line) with parameters $d = 65$ and $k = 40$) and with a correction for growth fraction of 0.95. (From Harley and Goldstein, 1980; after Levy et al., 1992, Figure 6, modified.) Source: Arino, O., Kimmel, M., and Webb, G.F. 1995. Mathematical modeling of the loss of telomere sequences. Journal of Theoretical Biology 177: 45–57. Figure 2a, page 53. Copyright: 1995 Academic Press Limited.

branching process model. The fit has been obtained using parameters $d = 65$ and $k = 40$.

7.2.5 Further work on telomere modeling

More recently, Olofsson and Kimmel (1999) and Olofsson (2000) considered models of telomere shortening involving the possibility of cell death, with the probability of the latter depending on cell type. These models exhibit a variety of limit behaviors, being the consequence of reducibility. The basic tools are the Tauberian theorems for probability generating functions.

7.3 Branching Random Walk with an Absorbing Barrier

In this section, we consider a different branching random walk model, leading to different dynamics (see the remarks at the end of this section). We consider a

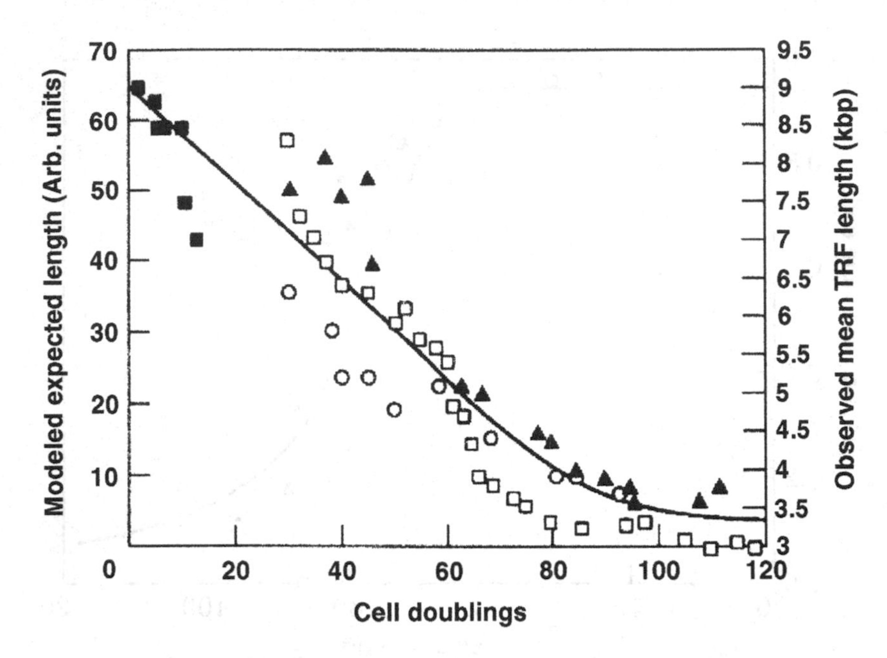

FIGURE 7.4. Mean length of terminal restriction fragments (TRFs) versus the number of cell doublings. Experimental data for a number of cultures of normal and transfected cells, up to the crisis time (from Counter et al., 1992), represented using differing symbols for different experiments, are compared to predictions using the Markov branching process model (continuous line) with parameters $d = 65$ and $k = 40$. Source: Arino, O., Kimmel, M., and Webb, G.F. 1995. Mathematical modeling of the loss of telomere sequences. Journal of Theoretical Biology 177: 45–57. Figure 3a, page 54. Copyright: 1995 Academic Press Limited.

population of abstract particles categorized into a denumerable quantity of types, denoted by $j = 0, 1, 2, \ldots$ and evolving according to the following rules:

1. The life spans of all particles are independent identically distributed exponential random variables with mean $1/\lambda$.
2. At the moment of death, a particle of type $j \geq 1$ produces two progeny particles each belonging to type $j + 1$ with probability b, to type $j - 1$ with probability d, and to type j with probability $1 - b - d$. However, a particle of type $j = 0$ produces two progeny of type 0.
3. The process is initiated at time $t = 0$ by a single particle of given type $i > 0$.

We consider the infinite vector $\mathbf{Z}(t) = (Z_0(t), Z_1(t), \ldots)$, where $Z_j(t)$ is the number of particles of type j at time t.

The main results obtained are as follows:

- Exact expressions for the expectations of the process, in the terms of modified Bessel functions
- Asymptotic expressions for the expectations, exponential modified by negative power terms

The distribution of $\mathbf{Z}(t)$ is determined by the probability generating functions (pgf's). Let $F_i(\mathbf{s}, t)$ denote the pgf of the infinite vector $\left(Z_0^{(i)}(t), Z_1^{(i)}(t), \ldots\right)$ of particle counts at time t, given that at time $t = 0$, there was exactly one particle of type i. As detailed in Kimmel and Stivers (1994), the pgf's satisfy the following infinite system of ordinary differential equations:

$$\frac{\partial F_0(\mathbf{s}, t)}{\partial t} = \lambda[F_0^2(\mathbf{s}, t) - F_0(\mathbf{s}, t)],$$

$$\frac{\partial F_i(\mathbf{s}, t)}{\partial t} = \lambda[f(F_{i-1}(\mathbf{s}, t), F_i(\mathbf{s}, t), F_{i+1}(\mathbf{s}, t)) - F_i(\mathbf{s}, t)], \quad i \geq 1, \tag{7.28}$$

where $f(s_{i-1}, s_i, s_{i+1}) = ds_{i-1}^2 + (1 - b - d)s_i^2 + bs_{i+1}^2$. The initial condition is $F_i(\mathbf{s}, 0) = s_i$.

Let $M_{ij}(t)$ denote the mean number of particles of type j at time t generated by a process starting with a single particle of type i at $t = 0$. Differentiating Eq. (7.28) with respect to s_j, we obtain

$$\frac{d}{dt} M_{ij}(t) = \lambda \left\{2dM_{i-1,j}(t) + [1 - 2(b + d)]M_{ij}(t) + 2bM_{i+1,j}(t)\right\}, \quad i \geq 1, \tag{7.29}$$

and $M_{0j} = e^{\lambda t}\delta_{0j}$. Equation (7.29) is a system of linear differential equations.

One way to solve Eq. (7.29) is to construct a generating function of the M_{ij}'s:

$$\mathcal{M}^j(u, t) = \sum_{i \geq 0} u^i M_{ij}(t), \quad j \geq 0.$$

Proceeding from the definition of $\mathcal{M}^j$, we obtain from Eq. (7.29),

$$\frac{d}{dt} \mathcal{M}^j(u, t) = \left[2(b + d) - \frac{2b}{u}\right] \lambda e^{\lambda t} \delta_{0j} + \lambda \left\{2du + [1 - 2(b + d)] + \frac{2b}{u}\right\} \mathcal{M}^j(u, t) - 2\lambda b M_{1j}(t).$$

If $j \neq 0$, then $\delta_{0j}(t) = 0$, so

$$\frac{d}{dt} \mathcal{M}^j(u, t) = A(u)\mathcal{M}^j(u, t) - 2b\lambda M_{1j}(t), \quad j \geq 1, \tag{7.30}$$

where

$$A(u) = \lambda \left[2du + 1 - 2(b + d) + \frac{2b}{u}\right].$$

Denoting the Laplace transform of $\mathcal{M}$ by $\hat{\mathcal{M}}$, we transform Eq. (7.30) with respect to t (Doetsch 1974):

$$p\hat{\mathcal{M}}^j(u, p) - \mathcal{M}^j(u, 0) = A(u)\hat{\mathcal{M}}^j(u, p) - 2b\lambda \hat{M}_{1j}(p).$$

Clearly, $M_{ij}(0) = \delta_{ij}$, so $\mathcal{M}^j(u, 0) = u^j$. Therefore, we obtain

$$\hat{\mathcal{M}}^j(u, p) = -\frac{u(u^j - 2b\lambda \hat{M}_{1j}(p))}{2d\lambda u^2 + ([1 - 2(b + d)]\lambda - p)u + 2b\lambda}. \tag{7.31}$$

$\hat{\mathcal{M}}^j(u, p)$ is analytic in p for any $u \in [0, 1)$. Therefore, if $\hat{u} \neq 0$ solves $u[A(u) - p] = 0$, then $\hat{u}$ must also be a root of $u^j - 2b\lambda\hat{M}_{1j}(p)$, which implies that $\hat{M}_{1j}(p) = \hat{u}^j/(2b\lambda)$. The roots of the denominator are

$$\hat{u}_i = \frac{p - \lambda[1-2(b+d)] + (-1)^i \sqrt{(\lambda[1-2(b+d)]-p)^2 - 16bd\lambda^2}}{4d\lambda}, \quad i = 1, 2. \tag{7.32}$$

Substituting $\hat{u} = \hat{u}_1$ into $\hat{M}_{1j}(p) = \hat{u}^j/(2b\lambda)$, we obtain $\lim_{p\to\infty} \hat{M}_{1j}(p) = 0$ (p real). The other root yields $\lim_{p\to\infty} \hat{M}_{1j}(p) = \infty$, inconsistent with the properties of the Laplace transform. Using $\hat{M}_{1j}(p) = \hat{u}_1^j/(2b\lambda)$, we obtain

$$\hat{M}_{1j}(p) = \frac{1}{2\lambda b}\hat{g}\left(\frac{p}{4\lambda d} - \frac{\lambda[1-2(b+d)]}{4\lambda d}\right), \tag{7.33}$$

where

$$\hat{g}(x) = \left\{x - \sqrt{x^2 - \frac{b}{d}}\right\}^j.$$

The counterimage of $\hat{M}_{1j}(p)$ is

$$M_{1j}(t) = \frac{je^{\lambda[1-2(b+d)]t}}{2b\lambda t}\left(\sqrt{\frac{b}{d}}\right)^j I_j(4\sqrt{bd}\lambda t), \tag{7.34}$$

where $I_j(z)$ is the modified Bessel function of order j (Abramowitz and Stegun 1972).

The following theorem describing the asymptotic behavior of $M_{1j}(t)$ was proved in Kimmel and Stivers (1994).

Theorem 27. *Suppose that $b < d$. Then,*

$$M_{1j}(t) \sim K_j \frac{e^{\lambda[1-2(\sqrt{b}-\sqrt{d})^2]t}}{t^{3/2}}, \quad \text{as } t \to \infty,$$

where

$$K_j = \frac{j\left(\sqrt{b/d}\right)^j}{4\lambda^{3/2}\sqrt{2\pi}\, b(bd)^{1/4}}, \quad j \geq 1,$$

and

$$\sum_{j\geq 1} M_{1j}(t) \sim K_S \frac{e^{\lambda[1-2(\sqrt{b}-\sqrt{d})^2]t}}{t^{3/2}} \quad \text{as } t \to \infty,$$

where

$$K_S = \frac{d\sqrt{\pi}}{4\lambda^{3/2}\sqrt{2\pi}(bd)^{1/4}(\sqrt{b}-\sqrt{d})^2}.$$

Moreover,

$$\sum_{j\geq 1} \frac{K_j}{K_S} = 1.$$

The consequence of Theorem 27 is that this branching random walk exhibits a property known as quasistationarity [mentioned in the context of Yaglom's Theorem (Theorem 7)]. We see that the entire population $\sum_{j \geq 0} M_{1j}(t)$ grows as $\exp(\lambda t)$. Because $\sum_{j \geq 1} M_{1j}(t)$ grows only as $t^{-3/2} e^{\lambda[1-2(\sqrt{b}-\sqrt{d})^2]t}$, this means that $M_{10}(t)$ grows as $\exp(\lambda t)$ [i.e., that type 0 is absorbing in the sense that $M_{10}(t)/\sum_{j \geq 0} M_{1j}(t) \rightarrow 1$, as $t \rightarrow \infty$]. However, $M_{1i}(t)/\sum_{j \geq 1} M_{1j}(t) \rightarrow K_i/K_S$ [i.e., the distribution of types conditional on nonabsorption tends to a limit (i.e., it reaches the quasistationary distribution)]. This quasistationary behavior of the random walk with an absorbing barrier is similar to that exhibited by the process of division-within-division (Section 7.5). In the next section, Theorem 27 will be applied to a model of unstable gene amplification, which may be considered an extension of the model of Section 3.6.

7.4 *Application*: A Model of Unstable Gene Amplification

This model is a time-continuous generalization of the random walk model from Kimmel and Axelrod (1990). No particular mechanism of gene amplification is assumed. It is only postulated that from one generation to another the number of gene copies on extrachromosomal elements may double or half. This model is based on the so-called Quantitative Shift Model described by Peterson (1984).

Hypotheses:

1. The life spans of cells are independent random variables distributed exponentially with mean $1/\lambda$.
2. **a.** The progeny of a cell having at least two gene copies may have twice as many gene copies per cell, the same number of gene copies per cell, or half as many gene copies per cell, with respective probabilities b, $1 - b - d$, and d.
   b. The progeny of a cell having a single copy of the gene may have two gene copies per cell, one gene copy per cell, or no gene copies per cell, with respective probabilities b, $1 - b - d$, and d.
   c. The progeny of a cell having no gene copies will also have no gene copies.

The constants b and d are the probabilities of gene amplification and deamplification. The asymptotic results of Theorem 27 apply directly if the following definition is used: A cell belongs to type $j \geq 1$ if it contains 2^{j-1} gene copies. A cell belongs to type $j = 0$ if it contains no gene copies.

Kimmel and Stivers (1994) employed this model to estimate probabilities of gene amplification and deamplification in cultured cells. Further analysis can be found, among others, in Bobrowski and Kimmel (1999).

7.5 Quasistationarity in a Branching Model of Division-Within-Division

Branching-within-branching occurs in various settings in cell and molecular biology. Examples include tightly regulated phenomena like replication of chromosomal DNA, but also processes in which the number of objects produced in each biological cell is a random variable. These are gene amplification in cancer cells, plasmid dynamics in bacteria, and proliferation of viral particles in host cells.

The general motivating idea is stability arising from selection superimposed on a random mechanism. We consider a set of large particles (biological cells), following a binary fission process. Each of the large particles is born containing a number of small particles (genes, viruses, organelles), which multiply or decay during the large particle's lifetime. The arising population of small particles is then split between the two progeny of the large particle and the process continues in each of them.

Let us suppose that the presence of at least one small particle is necessary to ensure the viability of the large particle. This can be due to a selection factor existing in the environment. One example of such selection factor is a cytotoxic drug, which eliminates cells (large particles) devoid of resistance genes (small particles), as in the gene amplification model of Section 7.4. We are interested in the behavior of the population of large particles surviving selection (i.e., large particles having at least one small particle inside).

We show that if the smaller particles follow a subcritical process, the number of smaller particles contained in a nonextinct large particle tends to a limit distribution. The result, in its present form (Kimmel 1997), depends on several detailed hypotheses, but these can be relaxed.

7.5.1 Definition of the process

Rules (schematically depicted in Fig. 7.5):

1. The population of large particles evolves according to a binary-fission time-continuous Markov branching process (Yule process) (i.e., each particle lives for a random time τ, exponential with parameter λ, and then splits into two progeny, each of which independently follows the same scenario).
2. Each large particle contains X small particles at its birth. Each of these proliferates, producing

$$Y^{(1)}, Y^{(2)}, \dots, Y^{(X)} \tag{7.35}$$

small particle progeny at the end of the large particle's lifetime.
3. Each of the $Y^{(k)}$ progeny of the initial kth small particle is independently split between the progeny of the large particle, so that large progeny 1 and 2 receive correspondingly $Y_1^{(k)}$ and $Y_2^{(k)}$ small progeny. The joint distributions of the pairs $(Y_1^{(k)}, Y_2^{(k)})$ are identical and independent for all (k), and symmetric in

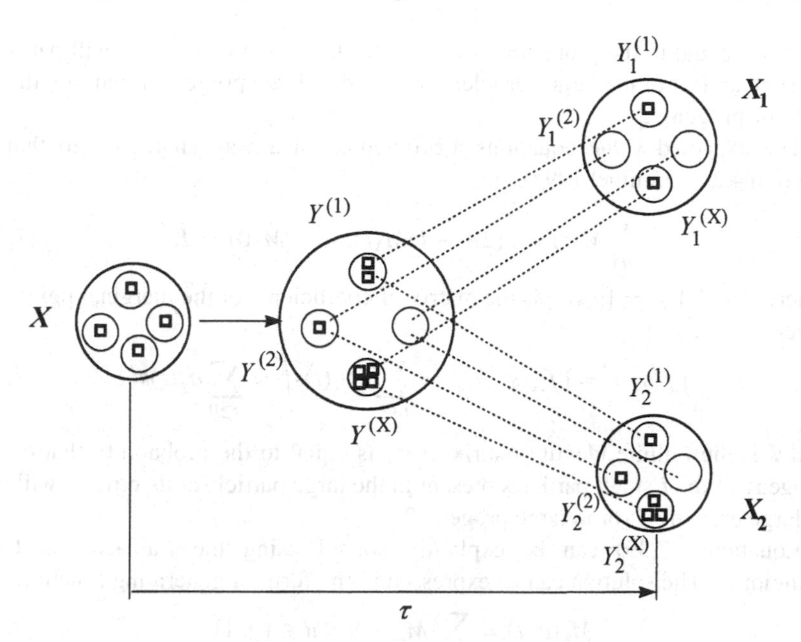

FIGURE 7.5. A large particle containing X small particles lives for a random time τ exponentially distributed with parameter λ and then splits into two progeny. During its lifetime, each of the X small particles proliferate, producing correspondingly $Y^{(1)}, Y^{(2)}, \ldots, Y^{(X)}$ small particles. Each of these $Y^{(k)}$'s is split independently among the two progeny of the large particle, so that large progeny 1 and 2 receive $\sum_{k=1}^{X} Y_1^{(k)}$ and $\sum_{k=1}^{X} Y_2^{(k)}$ small particles, respectively. The joint distributions of the pairs $(Y_1^{(k)}, Y_2^{(k)})$ are identical and symmetric.

$Y_1^{(k)}$ and $Y_2^{(k)}$. They are described by the joint probability generating function

$$f_{12}(s_1, s_2) = \mathrm{E}[s_1^{Y_1^{(1)}} s_2^{Y_2^{(1)}}]. \tag{7.36}$$

4. As a result, each of the large progeny receives the total of

$$X_1 = \sum_{k=1}^{X} Y_1^{(k)} \quad \text{and} \quad X_2 = \sum_{k=1}^{X} Y_2^{(k)} \tag{7.37}$$

small progeny particles.

The resulting branching process can be described as a Markov time-continuous process with a denumerable infinity of types of large particle. The large particle is of type i if it contains i copies of small particles at its birth. Let us denote the vector of counts of large particles of all types at time t by $Z(t) = [Z_0(t), Z_1(t), Z_2(t), \ldots]$ and the infinite matrix of expected values $M(t) = [M_{ij}(t)]$ by $M_{ij}(t) = \mathrm{E}[Z_j(t) | Z_i(0) = 1,\ Z_k(0) = 0,\ k \neq i]$.

Let us define coefficients $a_{nm}(i)$ using the expansion of the pgf of the sums in Eq. (7.37) given $X = i$,

$$[f_{12}(s_1, s_2)]^i = \sum_{n,m \geq 0} a_{nm}(i) s_1^n s_2^m. \tag{7.38}$$

$a_{nm}(i)$ is equal to the probability that among the progeny of the i small particles present at birth of the large particle, n will end in large progeny 1 and m will end in large progeny 2.

The expected value equations are obtained in a way analogous to that in Section 4.2.1 (Kimmel 1997):

$$\frac{d}{dt}M(t) = \lambda(2A - I)M(t), \qquad M(0) = I, \tag{7.39}$$

where $A = [A_{ij}] = [a_j(i)]$ is the matrix of coefficients of the marginal pgf of X_1 given $X = i$,

$$[f(s_1)]^i = [f_{12}(s_1, 1)]^i = \sum_{j,l\geq 0} a_{jl}(i)s_1^j = \sum_{j\geq 0} a_j(i)s_1^j, \tag{7.40}$$

and I is the infinite identity matrix. $a_j(i)$ is equal to the probability that of the progeny of the i small particles present in the large particle at its birth, j will end in large progeny 1 (or in large progeny 2).

Equations (7.39) can be explicitly solved using the Laplace transform techniques. The solution can be expressed in the form of generating function

$$M_k(u, t) = \sum_{l\geq 0} M_{kl}(t)u^l, \ u \in [0, 1]. \tag{7.41}$$

We obtain

$$M_k(u, t) = \sum_{j\geq 0} \frac{(2\lambda t)^j}{j!}[f_j(u)]^k e^{-\lambda t}, \ k \geq 0, \tag{7.42}$$

where $f_j(u)$ is the jth iterate of the marginal pgf of $Y_1^{(1)}$.

7.5.2 *Quasistationarity*

We begin with stating several facts concerning the Galton–Watson process with progeny pgf $f(u)$ (see Section 3.5.2).

If $f'(1-) < 1$ (the subcritical case), then, as $j \to \infty$,

$$\frac{f_j(u) - f_j(0)}{1 - f_j(0)} \to \mathcal{B}(u); \tag{7.43}$$

that is, conditionally on nonextinction, the process tends to a limit distribution, with pgf $\mathcal{B}(u)$ such that $\mathcal{B}(0) = 0$, $\mathcal{B}(1) = 1$ (cf. Athreya and Ney 1972, Corollary 1.8.1). This behavior is known as quasistationarity. Moreover, as $j \to \infty$,

$$f_j(u) - 1 \sim \rho^j Q(u), \tag{7.44}$$

where $\rho = f'(1-)$ and the function $Q(u)$ satisfies

$$\frac{Q(0) - Q(u)}{Q(0)} = \mathcal{B}(u), \tag{7.45}$$

with $Q(1) = 0$, $Q'(1-) = 1$, $Q(u) \leq 0$, and $Q(u)$ increasing for $u \in [0, 1]$.

Functions $\mathcal{B}(u)$ and $Q(u)$ are unique solutions of certain functional equations (Section 3.5.2). The following results are proved in Kimmel (1997).

Theorem 28. *Let us consider the process defined in Section 7.5.1 started by a large ancestor of type k and let* $\rho = f'(1-) < 1$. *Then, as* $t \to \infty$,

$$e^{\lambda t} - M_k(u, t) \sim -kQ(u)e^{(2\rho-1)\lambda t} \tag{7.46}$$

for all $k \geq 1$.

Corollary 3. *The expected frequencies* $\{\mu_{kl}(t),\ l \geq 1\}$ *of large particles of type l among the particles of nonzero type tend, as* $t \to \infty$, *to a limit distribution independent of k, characterized by the pgf* $\mathcal{B}(u)$.

7.5.3 Gene amplification

The process considered serves as another model of gene amplification. It is a direct generalization of the Galton–Watson process models of Section 3.6.2. Let us assume that large particles are cells and the small ones are copies of the gene conferring drug resistance located on extrachromosomal elements. Cells without any copies of the gene are eliminated by the drug (the selective agent). We accept the following specific hypotheses, similar to those in Kimmel and Axelrod (1990) and Kimmel and Stivers (1994):

- During the cell's lifetime, each extrachromosomal copy of the gene is successfully replicated with probability β, less than 1.
- The resulting two copies are segregated to the same progeny cell with probability α and to two different progeny cells with probability $1 - \alpha$. α may be called the probability of cosegregation and has been showed to be close to 1, using data from Kimmel and Axelrod (1990).

The above hypotheses yield

$$f_{12}(s_1, s_2) = \beta\left[(1-\alpha)s_1 s_2 + \frac{\alpha}{2}(s_1^2 + s_2^2)\right] + (1 - \beta) \tag{7.47}$$

and

$$f(u) = \frac{\beta\alpha}{2}u^2 + \beta(1-\alpha)u + \left(\frac{\beta\alpha}{2} + 1 - \beta\right), \tag{7.48}$$

with $\rho = f'(1-) = \beta < 1$. Therefore, our theorem and its corollary apply.

Qualitatively, all of the above experimental observations are explained by our results: The stable quasistationary distribution of gene copy count is predicted by the corollary.

If the type-0 cells are not removed by the drug, then the theorem proves that they dominate the population. Indeed, by the theorem the resistant cells grow as

$$\sum_{l \geq 1} M_{kl}(t) = M_k(1, t) - M_k(0, t) \sim -kQ(0)e^{(2\rho-1)\lambda t}, \quad \rho < 1, \tag{7.49}$$

whereas the entire population grows as $e^{\lambda t}$.

If $\rho > \frac{1}{2}$, then in the presence of the drug, resistant cells grow as $e^{(2\rho-1)\lambda t}$ (i.e., exponentially but slower than in the nonselective conditions).

7.6 Galton–Watson and Bellman–Harris Processes with Denumerably Many Types and Branching Random Walks

One of the questions arising when a multitype process is generalized to a denumerable infinity of types is, Which of the simple properties of the finite case remain valid? The answer is of importance for the applications, as it helps to decide what new properties are expected from a model if the constraint on the number of types is released. Applications in several sections in the present chapter demonstrate that, in general, a variety of asymptotic behaviors can be expected.

One of the articles devoted to this issue is Spătaru (1989). This article considers the extension of the multitype Galton–Watson process to countably many types indexed by $\mathbf{N}$. Let $Z_n = (Z_{n\alpha})$ be the vector of generation sizes; $Z_{n\alpha}$ is the number of α types in generation n ($\alpha \in \mathbf{N}$). Let $M = [M_{\alpha\beta}]$, where $M_{\alpha\beta} = \mathrm{E}(Z_{1\beta}|Z_{0\beta} = \delta_{\alpha\beta}, \beta \in \mathbf{N})$, be the mean matrix and let $f(s) = (f_\alpha(s))$, where, for $s \in C = [0, 1]^{\mathbf{N}}$, $f_\alpha(s) = \mathrm{E}(s^{Z_n}|Z_{0\beta} = \delta_{\alpha\beta}, \beta \in \mathbf{N})$. The author shows that the nonzero states are transient if M is irreducible and $f(s) \neq Ms$ for some $s \in C$. It is asserted that transience does not imply the property $\mathrm{P}(Z_n \to 0 \text{ or } |Z_n| \to \infty) = 1$, valid when the number of types is finite.

Let q denote the vector of extinction probabilities. If $f_n(s)$ denotes the vector-valued n-fold functional iterate of $f(s)$ [i.e., $f_{\alpha n}(s) = f_\alpha(f_{n-1}(s))$], then $q = \lim_{n\to\infty} f_n(0)$. For which s is it true that $f_n(s) \to q$? He shows that it is true if $s \leq q$ (coordinatewise), but not true for all $s \in C \backslash \{\mathbf{1}\}$, where $\mathbf{1} = (1, 1, \ldots)$, if $S = \{s \in C : s = f(s)\}$ has number of elements exceeding 2. If M is irreducible and f is not affine, then number of elements of S is equal to 2 and $q = \mathbf{1}$ or $q < \mathbf{1}$.

Another article on a related subject is by Moy (1967). A denumerable-type Galton–Watson process is considered, with mean progeny matrix $M = (m_{ij})$, where $m_{ij} = \mathrm{E}[Z_{n+1}(j)|Z_n = e_i]$. The principal role is played by the Perron–Frobenius root r of M, in this case equal to the radius of convergence of the power series $\sum_i M^i s^i$. The Perron–Frobenius root plays the role of the reciprocal of the Malthusian parameter. Two cases are possible: (I) $\sum_i M^i r^i$ finite and (II) $\sum_i M^i r^i$ is infinite. In Case I, there exist two strictly positive infinite sequences v and u, unique up to multiplicative constants, satisfying $rvM = v$ and $rMu = u$ (i.e., left and right eigenvectors corresponding to eigenvalue r^{-1}). In Case II, under an additional condition $\sum_i u(i)v(i) < \infty$ and if the process is supercritical (i.e., if $r^{-1} > 1$), $Z_n r^n$ converges in mean square to vY, where Y is a scalar random variable. In the remaining cases, $Z_n r^n$ converges to 0.

It seems that the conditions for asymptotic behavior of the supercritical process can be obtained as conclusions from the conditions for the general branching process of Appendix C.1. In the case of the branching process with denumerable-type space, these conditions seem to be a nontrivial extension of the positive regularity conditions sufficient in the finite multitype case (see Theorem 22). Indeed, the branching random walk of Section 7.3, conditional on not entering the 0-state, is a supercritical branching process. Its expected progeny matrix is irreducible in the sense of two arbitrary states communicating in a finite number of steps. However, as seen from Theorem 27, the asymptotics conditional on not entering the 0 state is exponential, modified by a negative power multiplier, and not a pure exponential. This means that the reproductive kernel of this process cannot be conservative, in the sense of condition (C.4), although a direct proof seems nontrivial.

Kesten (1989) proves a limit theorem for the rate of growth of a supercritical multitype branching process with countably many types. He proves, under appropriate conditions, that both the growth rate and the direction of growth in type space are essentially deterministic. The principal motivation for this work is to extend branching process theory to a problem arising in the study of random fractals (i.e., properties of the projections of random Cantor sets in d dimensions onto subspaces of small dimensions).

A large number of articles were written on the subject of branching random walks (i.e., denumerable-type branching processes with type-space transitions having the form of random walk). The typical problems considered include the rate of spread and growth of the branching random walk (Biggins 1977, 1995, 1997) and the Seneta–Heyde norming (Biggins and Kyprianou, 1996). A surprisingly small number of articles are devoted to branching random walks with restrictions, of the type considered in Section 7.3 or others. One example is Biggins et al. 1991, considering a supercritical branching random walk on the real line commencing with a single ancestor at the origin. All individuals reproduce according to the same law with mean family size $b > 1$. Each progeny is given an iid displacement from its parent with distribution F having negative mean and an exponentially decaying right tail [i.e., $\int_{-\infty}^{\infty} e^{st}\, dF(t) < \infty$ for some s]. The process is then attenuated by deleting all individuals below $-x$ and their descendants. Each remaining line of descent is just a random walk, starting at 0, with a barrier at $-x$, where $x > 0$. Results concerning the extinction probability and the expected population size depend on the parameter $h = \sup_\theta(-\log \int_{-\infty}^{\infty} e^{\theta t}\, dF(t))$. Specifically, if $b < e^h$, the probability that the process becomes extinct is 1, and if $b > e^h$, the probability of nonextinction is strictly positive. In the case $b < e^h$ and F nonlattice, the expected size of the total population, denoted by $f(x)$, satisfies $\lim_{x\to\infty} e^{-\alpha x} f(x) = C$, where α is the smallest positive solution of the equation $b \int_{-\infty}^{\infty} e^{\alpha t}\, dF(t) = 1$ and C is a positive constant which can be estimated.

7.6.1 Biological models with a denumerable infinity of types

An example of such application is the paper by Taïb (1993), where a branching model is proposed for the behavior of populations of the budding yeast *Saccharomyces cerevisiae* (also, see Gyllenberg 1986). Using the idea of branching processes counted by random characteristics (Section C.1.2), explicit expressions are obtained describing different aspects of the asymptotic composition of such populations. Using the author's words, "The main purpose of this note is to show that the branching process approach is an alternative to deterministic population models based on differential equation methods."

A complementary reading to the material of Section 7.2 is an article by Kowald (1997) which concerns the possible mechanisms for the regulation of telomere length. As mentioned in Section 7.2, because DNA polymerases can only synthesize a new DNA strand in the $5'$–$3'$ direction and need a primer that provides a free $3'$ end, the cellular replication machinery is unable to duplicate the $3'$ ends of linear chromosomes unless special mechanisms are operative. Although the telomeres seem to shorten continuously in human somatic cells because of the "end-replication" problem, it appears that telomere length is maintained in cancer cells, the germline, and unicellular organisms like yeast and *Tetrahymena* by a mechanism involving the enzyme telomerase, which elongates the $3'$ ends of telomeres. However, telomerase must be part of a more complicated mechanism to ensure that there is no net gain or loss of telomeric ends. Kowald (1997) described a simple theoretical model being, in essence, a denumerable-type branching process that can explain several experimental findings. The simulations show that (i) the proposed mechanism is able to maintain telomeres at a constant length, (ii) this length constancy is independent of the initial telomere length, (iii) mutations of the telomeric sequence lead to an elongation of telomeres, (iv) inhibition of telomerase causes telomeric shortening, and (v) it reproduces and explains the experimental result that the addition of oligonucleotides to the culture medium leads to an increase of telomere length. Although no formal mathematical analysis is carried out by Kowald (1997), the model may lead to interesting applications.

7.7 *Application*: Structured Cell Population Models

Structured population models describe proliferation of populations taking into account distributions of variables characterizing individuals. In the context of cell populations, examples of structural variables are cell mass, levels of biochemical constituents such as RNA or proteins, degree of cell maturation or differentiation, and so forth (Kimmel 1987). A frequently used method of modeling structured cell populations is by means of partial differential equations (PDE) of transport type. One of the most general models of this type was analyzed by Webb (1987, 1989). Another comprehensive reference is the book by Metz and Diekmann (1986). An alternative approach employs branching processes and more general stochastic processes (Arino and Kimmel 1993). Type space should be rich enough to accommodate a structure variable x, varying in a continuum [e.g., in an interval or another subset of the real line (Pakes and Trajstman 1985)]. This can be accomplished using general branching processes.

Another class of models describes the expected values of stochastic (branching) processes of cell proliferation. These models employ integral equations of renewal type, including type-transition laws in the kernel functions under the integral sign.

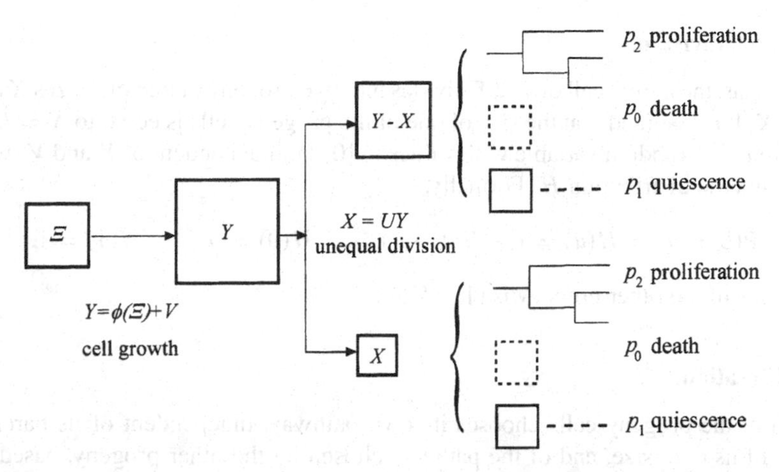

FIGURE 7.6. A schematic representation of the cell cycle model. A cell of size Ξ at its birth grows during a single generation to size $Y = \phi(\Xi) + V$, where ϕ is an increasing function and V is a non-negative random variable with cumulative distribution G. In mitosis, the cell divides into two daughters of unequal sizes X and $Y - X$ according to the rule $X = UY$, where U is a random variable on [0, 1] (independent of V) with a symmetric distribution H. Each of the daughter cells, independently, starts growing with probability p_2, dies with probability p_0, or becomes quiescent with probability p_1 ($p_0 + p_1 + p_2 = 1$).

Examples of such models are those by Tyson (1987), Tyson and Hannsgen (1985a, 1985b, 1986), Kimmel et al. (1984), Arino and Kimmel (1987), Arino et al. (1991), and others. Related ideas are explored in Alt and Tyson (1987) and Tyson et al. (1979).

7.7.1 *A model of unequal division and growth regulation in cell colonies*

The model introduced by Kimmel and Axelrod (1991) unifies features of a time-discrete Galton–Watson branching process and those of a deterministic model of cell cycle regulation introduced by Kimmel et al. (1984). It is a generalization of the example in Section 3.2. A schematic representation is depicted in Figure 7.6.

Cell growth.

A cell of size (mass, volume) X_0 at its birth grows during a single generation to size $Y = \phi(X_0) + V$, where V is a non-negative random variable with given cumulative distribution G. Function ϕ represents the size regulation mechanism; it is assumed nondecreasing, which means that progeny cells larger at birth are also larger at division. However, specific assumptions on ϕ ensure that any deviation from the average size, if present at the birth of the cell, decreases during cell growth. For mathematical simplicity, it is assumed that proliferating cells have identical lifetimes and that the lifetimes of the quiescent cells are infinite.

Unequal division.

In mitosis, the parent cell of size Y divides into two progeny of unequal sizes X and $Y - X$. It is assumed that the size of one of the progeny cells is equal to $X = UY$, where U is a random variable with values in [0, 1], independent of Y and V, with a symmetric distribution H. Formally,

$$\mathrm{P}\{U \leq u\} = H(u) = 1 - H(1-u), \qquad H(0) = 0, \qquad H(1) = 1.$$

The size of the other progeny is $(1 - U)Y$.

Proliferation.

Each of the progeny cells chooses its own pathway, independent of its parent's size, of its own size, and of the pathway chosen by the other progeny, based on a purely random rule. With probability p_2, the cell starts growing and initiates a pedigree, with probability p_0 it dies, or with probability p_1 it becomes quiescent (i.e., continues to exist without either growing or dying).

Independence.

Due to the assumed independence of cell death and quiescence from growth regulation and unequal division, one prediction of the present model is that the distribution of number of cells per colony does not depend on the birth size of the initial cell. In particular, this implies independence between number of cells within a colony and birth sizes of cells within the colony, at any time after the initiation. This is consistent with experimental observations (see Fig. 7.7 and Kimmel and Axelrod 1991).

Let us note that because of independence between cell proliferation, quiescence and death on one hand, and cell growth and unequal division on the other, the total count of proliferating and quiescent cells obey the laws of the Galton–Watson process in the example in Section 3.2. Therefore, we focus here on the size structure of the process.

Let

$$M_i(x, x_0) = \mathrm{E}[N_i(x, x_0)],$$
$$R_i(x, x_0) = \mathrm{E}[Q_i(x, x_0)]$$

denote the expected numbers of proliferating and quiescent cells with birth sizes not exceeding x, in the ith generation of the process started by a single cell with birth size x_0. These counting functions describe the cell size structure of the population.

Theorem 29. *Under suitable hypotheses (Kimmel and Axelrod 1991), the following recurrences are satisfied:*

$$M_{i+1}(x, x_0) = 2p_2 \int_0^\infty \int_x^\infty H\left(\frac{x}{y}\right) d_y M_i[\phi^{-1}(y - v), x_0]\, dG(v),$$
$$M_0(x, x_0) = 1(x - x_0) \tag{7.50}$$

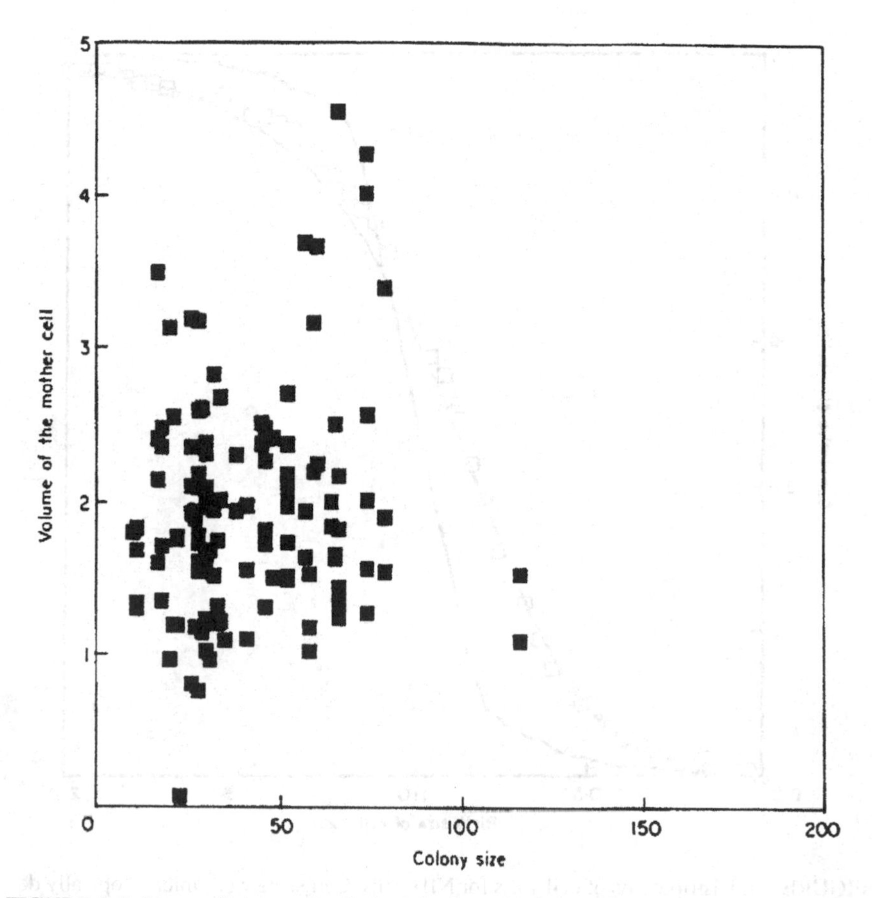

FIGURE 7.7. Size at division of cells in colonies with different number of cells. Colonies were grown for 4 days from single cells. Then, for each colony, cells were counted and the sizes of pairs of cells after division were determined. Up to three pairs of dividing cells per colony were recorded. The sum of the volumes of daughter cells is given as the volume of the mother cell. No dependence of cell size at division on colony size is apparent. Source: Kimmel, M. and D.E. Axelrod 1991. Unequal cell division, growth regulation and colony size of mammalian cells: A mathematical model and analysis of experimental data. Journal of Theoretical Biology 153: 157–180. Figure 2, page 161. Copyright: 1991 Academic Press Limited.

and

$$R_i(x, x_0) = \frac{p_1}{p_2} \sum_{j=1}^{i} M_j(x, x_0),$$

$$R_0(x, x_0) = 0. \tag{7.51}$$

Dynamics of cell size distributions.

The experimental data available for comparison with the model are the empirical distributions of cell size (understood as volume) in the same experimental system

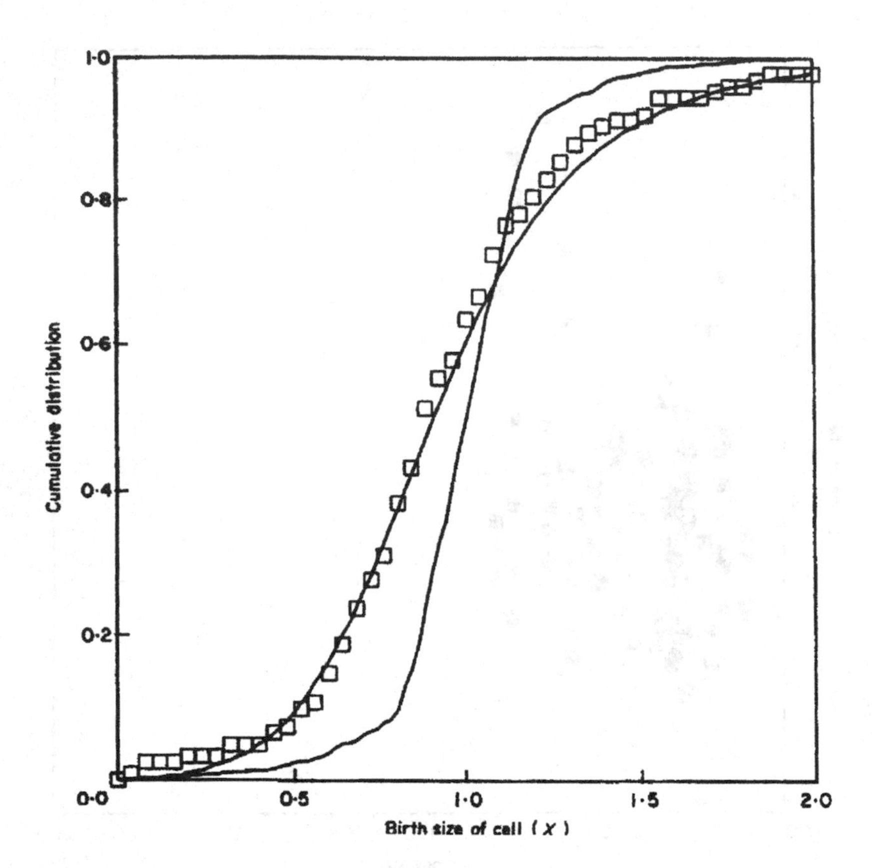

FIGURE 7.8. Distributions of cell sizes for NIH cells. Cell sizes were microscopically determined immediately after division. Symbols represent the observed cumulative frequency of the daughter cell sizes. Continuous lines are expected size distributions generated by the model, after one and eight generations, starting with a single founder cell of size 1 unit. The modeled distribution of cell sizes after eight generations closely resembles the empirically observed distribution. Source: Kimmel, M. and D.E. Axelrod 1991. Unequal cell division, growth regulation and colony size of mammalian cells: A mathematical model and analysis of experimental data. Journal of Theoretical Biology 153: 157–180. Figure 4, page 164. Copyright: 1991 Academic Press Limited.

as presented in the chapter concerning the Galton–Watson process. Figure 7.8 depicts the cumulative distribution of sizes of the measured NIH progeny cells. The corresponding distribution of the NIH(*ras*) cells is indistinguishable.

The question to answer by mathematical modeling is the following: Using the empirical distribution H of inequality of cell division and a mathematical form of the growth function ϕ, is it possible to reproduce the observed size distribution at the end of the experiment?

Figure 7.8 also depicts the evolution of distributions of cell size modeled using Eqs. (7.50) and (7.51). To obtain cumulative distributions, the counting functions have been normed. If it is assumed that the size of the founder cell of the colony was $x_0 = 1$, the modeled colony size distribution at generation 8 (end of experiment) is close both to the limit distribution and to the empirical distribution. If the founder cell size is assumed very small or very large, the convergence to the limit distribution is still satisfactory within 10 generations (not shown). The fit provided by the model is satisfactory for cell size distributions as well as colony size distributions (Section 3.2).

7.7.2 *Cell cycle model with cell size control, unequal division of cells, and two cell types*

Among many laboratory and mathematical models of structured cell populations, the system introduced by Sennerstam (1988) stands out because it encompasses almost all features encountered in such systems. Also, it gave rise to a series of studies ranging from laboratory investigation, through theorizing and computer modeling, to mathematically advanced models in the form of renewal equations and general branching processes.

Similar to the model in the preceding section, Sennerstam's (1988) studies were motivated by the observation made in the 1960s that the partition of mass to daughter cells at mitosis is asymmetric. Furthermore, it was suggested that such an unequal distribution of metabolic constituents at mitosis contributes to the dispersion of cell generation times and cell masses in a population [for other examples of unequal division, see Birky and Skavaril (1984), Czerniak et al. (1992), Kotenko et al. (1987) and Lapidus (1984)]. Various theories were proposed concerning the mechanisms of regulation of generation time and cell growth rate, given the cell's birth mass and other factors (among them Darzynkiewicz et al. 1979, 1982, Cooper 1984, Kimmel et al. 1984).

In Sennerstam (1988), cultured PCC3 embryonal carcinoma (EC) cells were studied in order to evaluate their protein content. There exists a considerable intraclonal intermitotic time heterogeneity found in undifferentiated PCC3 EC cells. It was concluded that the postmitotic difference in mass (protein) between sister cell pairs has an influence upon the variation in cell cycle time duration when comparing sister cell pairs. This offered an explanation for the randomly distributed difference repeatedly found between sister cell generation times. In spite of this, there was no correlation seen between the mass difference found between sister cell pairs postmitotically and the mass of the mother cell.

In subsequent articles, Sennerstam and Strömberg (1988) reported the discovery of an intraclonal bimodal-like cell cycle time variation within the multipotent embryonal carcinoma (EC) PCC3 N/1 line. The variability was found to be localized in the G_1 period. Furthermore, an inverse relation between cell mass and cell generation time was found in the cell system analyzed. It was suggested that the bimodal intraclonal time variability previously reported was attributable to an intraclonal shift between two types of cell-growth-rate cycles.

To explain the findings, Sennerstam and Strömberg (1995, 1996) used the so-called continuum model (Cooper 1979, 1984, 1991). The model is based on the idea that DNA replication and cell growth are two loosely coupled subcycles.

After division (generally asymmetric), a cell proceeds through the G_1 phase until it reaches a checkpoint characterized by a threshold mass. At this moment, the DNA synthesis is triggered and the time to division and further mass increase (or a growth rate) are determined. The growth continues after division at the same rate and so forth. In this way, the division cycle (from one cell division to another) is only partly coupled with the growth cycle, because adjustments to the growth rate are made only at the G_1/S boundary checkpoints. Thresholds and rates have stochastic components and, consequently, the mass-at-division regulation is not perfect [also, see Kuczek (1984)].

Sennerstam's measurements described above were used by Kimmel and Arino (1991) to build a mathematical model, equivalent to expected value equations for a branching process. An extremely simplified version was already mentioned in Section 6.3. The model takes into account cell size regulation (cells grow between divisions, at certain mass they decide to divide), unequal division (some cell constituents do not split equally between progeny cells), and differentiation (cells switch off/on some of their genes, to specialize in a required direction).

The following detailed observations were listed in Sennerstam (1988) describing growth characteristics of immortalized embryonic cells:

- Using mitotic detachment technique, it was established that the coefficient of variation of the mass of progeny cells exceeded the coefficient of variation of parent cells by about 4%, that is,

$$cv_{parent\ mass}/cv_{progeny\ mass} \cong 1.04.$$

- Using time-lapse measurements, the distributions of the generation times of related cells were determined. The indexation of generation (interdivision) times and other variables describing cell pedigrees is explained by the following diagram:

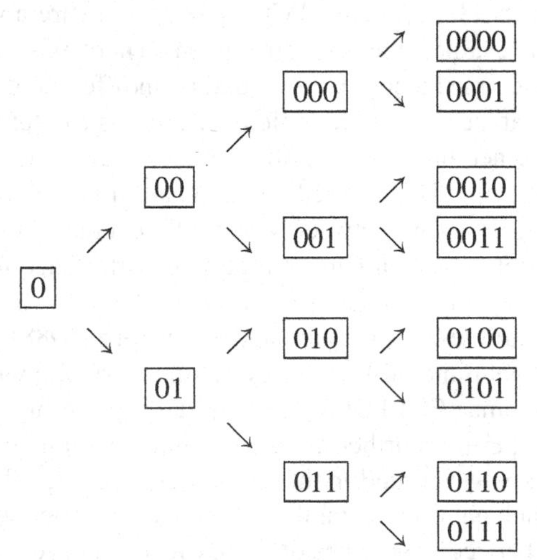

α curve $= f_{T_0}(\tau)$, the distribution of cell life lengths, was found to be bimodal.

β_1 curve $= f_{|T_{00}-T_{01}|}(\tau)$, the distribution of differences of sib cells life lengths, was found to be unimodal.
β_2 curve $= f_{|T_{000}-T_{011}|}(\tau)$, the distribution of differences of the first cousin cells life lengths, was found to be bimodal.
β_3 curve $= f_{|T_{0000}-T_{0111}|}(\tau)$, the distribution of differences of the second cousin cells life lengths, was found to be unimodal.

- Furthermore, correlation coefficients between generation times of related cells were estimated:

  Parent–progeny, $\rho_{T_0,T_{00}} = 0.77$
  Sib–sib, $\rho_{T_{00},T_{01}} = 0.95$
  Cousin–cousin, $\rho_{T_{000},T_{011}} = 0.41$

Kimmel and Arino (1991) proposed the following model, also based on Cooper's (1984) continuum hypothesis, to explain the observations listed above:

- There exist two types of cells: type $\boxed{1}$, smaller and faster cycling; type $\boxed{2}$, larger and slower cycling.
- Growth of cell mass between divisions proceeds at a constant rate r. Specifically,

  A cell of type i and initial mass y, grows in the G_1-phase to a random threshold size

$$w_i \sim h_i(\cdot), \qquad w_1 \overset{(d)}{<} w_2,$$

  where the stochastic inequality between w_1 and w_2 is defined as the relation between their tail distributions (i.e., $P[w_1 > x] < P[w_2 > x]$);
  Then, it continues through phases S+G_2+M for a constant time τ (that is, the total duration of the cell cycle is equal to

$$T = \frac{w_i - y}{r} + \tau.$$

  It grows to the predivision mass x,

$$x = w_i + r\tau.$$

- Switching between types: At the checkpoint on the G_1/S boundary, it is decided whether the type of progeny (both) is the same as the parent:

$$\Pr[i \to j] = p_{ij}.$$

This is the "supramitotic regulation" (i.e., decisions are made at a checkpoint inside the division cycle).

- Unequal division: Parental cell of mass x_0 divides into progeny of masses y_{00} and y_{01}. Asymmetry of division can be represented by multiplication of x_0 by a random variable u_0, with distribution with support in [0, 1], as represented in the following diagram:

$$\boxed{\text{i}|\text{j}}_0 \quad \begin{array}{ll} \nearrow \boxed{\text{j}|\cdots}_{00} & y_{00} = u_0 x_0 \\ \searrow \boxed{\text{j}|\cdots}_{01} & y_{01} = (1 - u_0) x_0 \end{array},$$

$$u_0 = f_0(u), \quad f_0\,(u) = f_0(1 - u).$$

The model explains the observations of Sennerstam (1988). Let us assume that the cell population is in the state of asynchronous exponential growth; that is,

$$\begin{pmatrix} N_1(t) \\ N_2(t) \end{pmatrix} = C \begin{pmatrix} \tilde{p}_1 \\ \tilde{p}_2 \end{pmatrix} \exp(\lambda t),$$

where $N_i(t)$ is the number of cells of type i at time t. To address the bimodality of the α curves, let us suppose that a cell of type i is the progeny of a cell of type j. This occurs with probability

$$\tilde{p}_j p_{ji}.$$

Then,

$$T_0|i \sim \tau + \frac{w_i - y_j}{r} = \tau + \frac{w_i - (w_j + r\tau)u}{r}.$$

This latter, under equal division, $u = 1/2$, reduces to

$$\frac{\tau}{2} + w_i - \frac{w_j}{2}.$$

Now, let us assume that the distribution of u is tightly concentrated around $1/2$. If in addition, $\tilde{p}_1 \cong \tilde{p}_2 \cong 1/2$, and p_{12} and p_{21} are small, then the two dominating modes of the distribution of random variable T_0 are approximately located at

$$\frac{\tau}{2} + \frac{w_i}{2}, \qquad i = 1, 2.$$

Unimodality of distributions of differences of life lengths of sib cells (the β_1-curves) follows because

$$T_{00} - T_{01} = \left[\tau + \frac{w_{00} + (w_0 + r\tau)u_0}{r}\right] - \left[\tau + \frac{w_{01} + (w_0 + r\tau)(1 - u_0)}{r}\right],$$

which, under $u = 1/2$, is equal to

$$\frac{w_{00} - w_{01}}{r},$$

so that $|T_{00} - T_{01}|$ has the only mode at zero.

Bimodality of first-cousin life length difference distributions (the β_2 curves) follows because

$$T_{000} - T_{010} = \frac{1}{r}\left[(w_{000} - w_{010}) - r\tau(u_{00} - u_{01}) - (w_{00}u_{00} - w_{01}u_{01})\right].$$

Again, under $u = 1/2$, this is equal to

$$\frac{1}{r}\left[\underbrace{(w_{000} - w_{010})}_{w_i - w_j} - \frac{1}{2}\underbrace{(w_{00} - w_{01})}_{\text{same distribution}}\right],$$

so the distribution of $|T_{000} - T_{010}|$ has one mode at 0 and another (smaller) at $|w_1 - w_2|$.

In addition to the above, Kimmel and Arino (1991) carried out computations of the correlations of related cells, under assumption $p_{11} = p_{22}$, with all other parameters fitted to data. The conclusions are as follows:

- The parent–progeny correlation is negative if p_{11} is small (frequent switching of type) and positive if p_{11} is large.
- The sib–sib correlation is more or less stable (same w).
- The cousin–cousin correlation is large if p_{11} is large (type rarely changed) and if p_{11} is small (type likely to be the same for each second generation).

An important theoretical problem concerns the dynamics of cell proliferation in this case: How to reconcile the division cycle with the growth cycle and with unequal division and the random decisions to switch cell type (these latter assumed to be taken at the G_1/S checkpoint)? It seems convenient to introduce four types of cells, indexed by pairs $(i, j)_{i,j=1,2}$:

[i | j] = type i that decided on progeny type j.

Transitions reduced to "decision taken at birth" are depicted in the following diagram:

p_{11} (loop on [1 | 1])

[1 | 1] $\xrightarrow{p_{12}}$ [1 | 2]

$\uparrow p_{11}$ $\nearrow p_{12}$ $\swarrow p_{21}$ $p_{22} \downarrow$

[2 | 1] $\xleftarrow{p_{21}}$ [2 | 2]

p_{22} (loop on [2 | 2])

Using these transitions allows writing straightforward balance equations for expected densities of flow rates between types. Suppose that

$$n_{ij}(t, y)\, dt\, dy$$

is the expected flow of (i, j) progeny of size in $(y, y + dy)$ into the growth phase, in the time interval $(t, t + dt)$; then,

$$n(t, y) = 2r \int f(x, y)H(x - r\tau) \int n[t - (\tau + \sigma),\ x - r(\tau + \sigma)]\, d\sigma dx,$$

where

$$n(t, y) = \begin{pmatrix} n_{11}(t, y) \\ n_{12}(t, y) \\ n_{21}(t, y) \\ n_{22}(t, y) \end{pmatrix},$$

$$H(w) = \begin{pmatrix} p_{11}h_1(w) & 0 & p_{11}h_2(w) & 0 \\ p_{12}h_1(w) & 0 & p_{12}h_2(w) & 0 \\ 0 & p_{21}h_1(w) & 0 & p_{21}h_2(w) \\ 0 & p_{22}h_1(w) & 0 & p_{22}h_2(w) \end{pmatrix}.$$

As demonstrated in Kimmel and Arino (1991), this evolution equation generates a semigroup of operators on $X = L^1(E)$, the space of functions integrable on E [i.e., such that $n(t) = \int_{y \in E} n(t, y)\, dy < \infty$], where the set E of admissible sizes of progeny is defined by specific assumptions on distributions $h_1(w)$ and $h_2(w)$:

$$G(t) : X \ni n_0 \longrightarrow n_t.$$

The main mathematical problem is to show that asynchronous exponential growth exists. It is sufficient to show that spectrum of $G(t)$ has a dominating eigenvalue $\exp(\lambda t)$ and that the corresponding generalized eigenspace is one dimensional. This is true because $G(t)$ is eventually compact. The projection of solution n_t on the generalized eigenspace dominates all other solutions and yields

$$n(t) \sim \exp(\lambda t)$$

and

$$N(t) \sim \exp(\lambda t)$$

as desired.

Alexandersson (1999) proposed a largely equivalent description in the form of a general branching process (Section C.1). The process has type space

$$\{11, 12, 21, 22\} \times \{R_+\}$$

with reproduction measure determined by the transition rules above. Finding Malthusian parameter for the process is equivalent to solving the characteristic equation for the dominant eigenvalue in the model by Kimmel and Arino (1991). Then, the problem is to demonstrate conservativeness of the reproduction measure. This is in some sense equivalent to proving results concerning the eigenspace corresponding to the dominant eigenvalue of the semigroup $G(t)$.

Alexandersson (1999) considers various versions of the regulation mechanism of the cell growth, some of them involving variable growth rates. These who venture to read both Kimmel and Arino (1991) and Alexandersson (1999) will see that the general branching process methodology makes the modeling process conceptually more straightforward. A discussion of Alexandersson (1999) is provided in Appendix C (Section C.2).

7.8 *Application*: Yule's Evolutionary Process

We paraphrase the branching process model of evolution of Yule (1924), as cited in Harris (1963). The model concerns the evolution of two basic taxonomic units, species and genera, within a single family. The following assumptions define the model:

1. Two types of objects are considered.
   a. *Species*. This is the smallest taxonomic unit. Different species are reproducively separated (i.e., if individuals belonging to different species are crossed, they do not produce fertile progeny).
   b. *Genus*. Species are grouped in genera. The biological distance separating different genera is larger than that separating different species.
2. Within a single genus, the collection of species evolves as an age-dependent branching process with exponential lifetime distributions with parameter λ and the pgf of the number of progeny equal to $h(s) = s^2$ (i.e., each speciation event produces exactly two new species).
3. The collection of genera evolves as an age-dependent branching process with exponential lifetime distributions with parameter γ. However, at each ramification, a new genus evolves which has exactly one species and the old genus continues unchanged. This asymmetry is caused by the fact that a new genus arises from a major evolutionary rearrangement within a single species of the old genus.

The object defined is a sort of a "branching process within a branching process" (Fig. 7.9).

Let $N(t)$ denote the number of genera existing at time t and let $M_i(t)$ denote the number of species existing at t in the ith genus. Then, the process can be defined as the vector

$$\mathcal{Z}(t) = \{M_1(t), M_2(t), \ldots, M_{N(t)}(t)\}, \quad t \geq 0.$$

Finding a comprehensive description for $\mathcal{Z}(t)$ is quite complicated because it is an age-dependent branching process with infinitely many particle types. However, we are interested in answering a very particular question regarding the process: What is the rate of evolution of new genera compared to evolution of new species, as measured by the ratio of γ to λ?

Let us note (see Fig. 7.9) that a high γ/λ ratio yields a relatively high frequency of monotypes, (i.e., genera with one species). Therefore, by examining the population-based proportion of monotypes at given time t, it seems possible to estimate the γ/λ ratio. This gives a chance to infer about the dynamics of the evolutionary process without actually observing it. It is particularly important in view of patchiness and discontinuity of paleontological evidence.

We develop a model with only two classes of genera: class 1 containing monotypes and class 2 genera with more than one species. The number, at time t, of genera of class i is denoted by $Z_i(t)$, $t = 1, 2$. The joint pgf of the pair

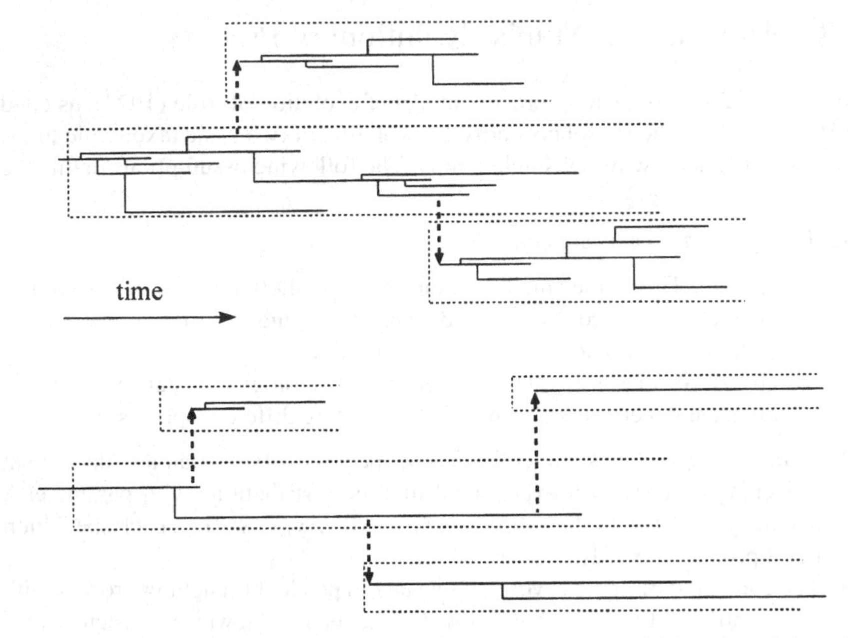

FIGURE 7.9. Two sample paths of the Yule's process: low and high value of the γ/λ ratio. Branching of species is represented by continuous lines. Boundaries of genera are represented by dotted line "tubes" and branching of genera is represented by arrows.

$(Z_1(t), Z_2(t))$ in the process started by a single class i genus is denoted by $F_i(s_1, s_2; t)$. We show the process is a two-type Markov age-dependent process.

The lifetime distribution of a class 2 genus is $G_2(\tau) = 1 - e^{-\gamma\tau}$. The joint pgf of the class 1 and 2 progeny of such genus is $h_2(s_1, s_2) = s_1 s_2$.

A class 1 (monotype) genus transforms into a class 2 genus after a time τ' distributed exponentially with parameter λ, because of a speciation event; independently, it splits into two genera after a time τ'' distributed exponentially with parameter γ. The minimum of these two times is τ distributed exponentially with parameter $\lambda + \gamma$. If $\tau' < \tau''$, which happens with probability $\lambda/(\lambda + \gamma)$, then the "progeny" pgf is s_2. Otherwise, it is s_1^2. Eventually, $G_1(\tau) = 1 - e^{-(\lambda+\gamma)\tau}$ and

$$h_1(s_1, s_2) = \frac{\lambda}{\lambda + \gamma} s_2 + \frac{\gamma}{\lambda + \gamma} s_1^2.$$

For a two-type age-dependent branching process, the pgf equations are a straightforward generalization of Eq. 4.6:

$$\frac{\partial F_i(s; t)}{\partial t} = \lambda_i \{h_i[F_1(s, t), F_2(s, t)] - F_i(s, t)\}, \quad t \geq 0, \; F_i(s; 0) = s_i, \; i = 1, 2. \tag{7.52}$$

This yields, in our case,

$$\frac{\partial F_1}{\partial t} = \gamma F_1^2 + \lambda F_2 - (\lambda + \gamma) F_1, \tag{7.53}$$

$$\frac{\partial F_2}{\partial t} = \gamma F_1 F_2 - \gamma F_2. \tag{7.54}$$

Using methods similar as in the application to clonal resistance theory and, in particular, a variant of Theorem 12, an explicit solution of the above system is obtained:

$$F_2 = \frac{s_2(\lambda+\gamma)e^{-\gamma t}}{[\lambda(1-s_2)+\gamma(1-s_1)] + s_2(\lambda+\gamma)e^{-\gamma t} + (s_1-s_2)\gamma e^{-(\gamma+\lambda)t}},$$

$$F_1 = \left(1 + \frac{s_1 - s_2}{s_2} e^{-\lambda t}\right) F_2.$$

Suppose the process (i.e., given family of genera) was started at time 0 by a monotypic genus. Then, the expected numbers of monotypic and polytypic genera at time t are given by

$$M_{11}(t) = \left.\frac{\partial F_1(s;t)}{\partial s_1}\right|_{s=(1,1)} = \frac{\gamma}{\lambda+\gamma} e^{\gamma t} + \frac{\lambda}{\lambda+\gamma} e^{-\lambda t}, \tag{7.55}$$

$$M_{12}(t) = \left.\frac{\partial F_1(s;t)}{\partial s_2}\right|_{s=(1,1)} = \frac{\lambda}{\lambda+\gamma} (e^{\gamma t} - e^{-\lambda t}). \tag{7.56}$$

Eventually, the expected proportion of monotypic genera in the family which is old enough is equal to

$$p = \lim_{t\to\infty} \frac{M_{11}(t)}{M_{11}(t) + M_{12}(t)} = \frac{\gamma}{\lambda+\gamma}. \tag{7.57}$$

In the book of Harris (1963), an example is quoted of a family of beetles with 627 genera comprising 9997 species, $p = 34.29\%$ of the genera being monotypes. From this, it can be estimated that $\lambda/\gamma = 1.9$.

CHAPTER 8

References

Abkowitz, J.L., Catlin, S.N. and Guttorp, P. 1996. Evidence that hematopoiesis may be a stochastic process in vivo. Nature Medicine 2: 190-197.

Abramowitz, M. and Stegun, I.A. (eds.) 1972. Handbook of Mathematical Functions with Formulas, Graphs, and Mathematical Tables. 10th printing. U.S. Government Printing Office, Washington, D.C.

Alexandersson, M. 1999. Branching processes and cell populations. Ph.D. thesis. Department of Mathematical Statistics, Chalmers University, Göteborg, Sweden.

Alt, W. and Tyson, J.J. 1987. A stochastic model of cell division (with application to fission yeast). Mathematical Biosciences 83: 1-29.

Angerer W.P. 2001. An explicit representation of the Luria–Delbrück distribution. Journal of Mathematical Biology 42: 145-174.

Arino, O. and Kimmel, M. 1987. Asymptotic analysis of a cell cycle model based on unequal division. SIAM Journal of Applied Mathematics 47: 128-145.

Arino, O. and M. Kimmel. 1991. Asymptotic behavior of nonlinear semigroup describing a model of selective cell growth regulation. Journal of Mathematical Biology 29: 289-314.

Arino, O. and Kimmel, M. 1993. Comparison of approaches to modeling of cell population dynamics. SIAM Journal of Applied Mathematics 53: 1480-1504.

Arino O., Kimmel M. and Webb G.F. 1995. Mathematical modeling of the loss of telomere sequences. Journal of Theoretical Biology 177: 45-57.

Arino, O., Kimmel, M. and Zerner, M. 1991. Analysis of a cell population model with unequal division and random transition. In: Mathematical Population Dynamics (Arino, O., Axelrod, D.E. and Kimmel, M. eds.). Marcel Dekker, New York, pp. 3-12.

Arking, R. 1998. Biology of Aging. Sinauer, Sunderland, MA.

Asmussen, S. and Hering, H. 1983. Branching Processes. Birkhauser, Boston, MA.

Asteris, G. and Sarkar, S. 1996. Bayesian procedures for the estimation of mutation rates from fluctuation experiments. Genetics 142: 313-326.

Athreya, K.B. and Ney, P.E. 1972. Branching Processes. Springer-Verlag, Berlin.

Axelrod, D.E. and Kuczek, T. 1989. Clonal heterogeneity in populations of normal and tumor cells. Computers and Mathematics with Applications 18: 871-881.

Axelrod D.E., Baggerly, K.A. and Kimmel, M. 1994. Gene amplification by unequal sister chromatid exchange: Probabilistic modeling and analysis of drug resistance data. Journal of Theoretical Biology 168: 151-159.

Axelrod, D.E., Gusev, Y. and Gamel, J.W. 1997. Ras-oncogene transformed and non-transformed cell population are each heterogeneous but respond differently to the chemotherapeutic drug cytosine arabinoside (Ara-C). Cancer Chemotherapy and Pharmacology 39: 445-451.

Axelrod, D.E., Gusev, Y. and Kuczek, T. 1993. Persistence of cell cycle times over many generations as determined by heritability of colony sizes of ras oncogene-transformed and non-transformed cells. Cell Proliferation 26:235-249.

Axelrod, D. E., Haider, F. R. and Tate, A. C. 1986. Distribution of interdivisional times in proliferating and differentiating Friend murine erythroleukaemia cells. Cell and Tissue Kinetics 19: 547-556.

Baggerly, K.A. and Kimmel, M. 1995. Emergence of stable DNA repeats from random sequences under unequal sister chromatid exchange. In: Proceedings of the 1st World Congress of Nonlinear Analysts, Tampa, Florida, August 1992 (Lakshmikantham, V., ed.). Walter de Gruyter, Berlin, pp. 3409-3418.

Bat, O., Kimmel, M. and Axelrod, D.E. 1997. Computer simulation of expansions of DNA triplet repeats in the fragile X syndrome and Huntington's disease. Journal of Theoretical Biology 188: 53-67.

Beaudry, A.A. and Joyce, G.F. 1992. Directed evolution of an RNA enzyme. Science 257: 635-641.

Berndtsson, B. and Jagers, P. 1979. Exponential growth of a branching process usually implies stable age distribution. Journal of Applied Probability 16: 651-656.

Bertuzzi, A. and Gandolfi, A. 1983. Recent views on the cell cycle structure. Bulletin of Mathematical Biology 45: 605-616.

Bertuzzi, A., Gandolfi, A., Giovenco, M. and Adelaide, M. 1981. Mathematical models of the cell cycle with a view to tumor studies. Mathematical Biosciences 53: 159-188.

Biggins, J.D. 1977. Chernoff's theorem in the branching random walk. Journal of Applied Probability 14: 630-636.

Biggins, J.D. 1995. The growth and spread of the general branching random walk. Annals of Applied Probability 5: 1008-1024.

Biggins, J.D. 1997. How fast does a general branching random walk spread? In, Classical and Modern Branching Processes (Jagers, P. and Athreya, K., eds.)

The IMA Volumes in Mathematics and Its Applications, 84. Springer-Verlag, Berlin, pp. 19-39.

Biggins, J.D. and Kyprianou, A.E. 1996. Branching random walk: Seneta–Heyde norming. In: Trees (Chauvin, B., Cohen, S. and Rouault, A. eds.). Progress in Probability, 40. Birkhäuser, Basel, pp. 31-49.

Biggins, J.D., Lubachevsky, B.D., Shwartz, A. and Weiss, A. 1991. A branching random walk with a barrier. The Annals of Applied Probability 1: 573-581.

Birky, C.W. and Skavaril, R.V. 1984. Random patitioning of cytoplasmic organelles at cell division: The effect of organelle and cell volume. Journal of Thoretical Biology 106: 441-447.

Blackburn, E.H. 1991. Structure and function of telomeres. Nature 350: 569-573.

Bobrowski, A. and Kimmel, M. 1999. Asymptotic behaviour of an operator exponential related to branching random walk models of DNA repeats. Journal of Biological Systems 7: 33-43.

Borovkov, K.A. and Vatutin, V.A. 1977. Reduced critical branching processes in random environment. Stochastic Processes and Their Applications 77: 225-240.

Breiman, L. 1968. Probability. Addison-Wesley, Reading, MA.

Brooks, R.F., Bennett, D.C. and Smith, J.A. 1980. Mammalian cell cycles need two random transitions. Cell 19: 493-504.

Brown, P.C., Beverly, S.M. and R.T. Schimke, R.T. 1981. Relationship of amplified dihydrofolate reductase genes to double minute chromosomes in unstably resistant mouse fibroblasts cell lines. Molecular and Cellular Biology 1: 1077-1083.

Caskey, C.T., Pizutti, A., Fu, Y.-H., Fenwick, R.G., Jr. and Nelson, D.L. 1992. Triplet repeat mutations in human disease. Science 256: 784-789.

Chinnery P.F. and Turnbull, D.M. 1999. Mitochondrial DNA and disease. Lancet 354: 17-21.

Ciampi, A., Kates, L., Buick, R., Kriukov, Y. and Till, J. E. 1986. Multi-type Galton–Watson process as a model for proliferating human tumour cell populations derived from stem cells: Estimation of stem cell self-renewal probabilities in human ovarian carcinomas. Cell Tissue Kinetics 19: 129-140.

Cohn, H. and Klebaner, F. 1986. Geometric rate of growth in Markov Chains with applications to population-size-dependent models with dependent offspring. Stochastic Analysis and Applications 4: 283-308.

Coldman, A. J. 1987. Modeling resistance to cancer chemotherapeutic agents. In, Cancer Modeling (Thompson, J.R. and Brown, B.W., eds.). Marcel Dekker, Inc. New York, pp. 315-364.

Coldman, A. J. and Goldie, A.J. 1983. A model for the resistance of tumor cells to cancer chemotherapeutic agents. Mathematical Biosciences 65: 291-307.

Coldman, A. J. and Goldie, J. H. 1985. Role of mathematical modeling in protocol formulation in cancer chemotherapy. Cancer Treatment Reports 69: 1041-1048.

Coldman, A. J. and Goldie, J. H. 1986. A stochastic model for the origin and treatment of tumors containing drug resistant cells. Bulletin of Mathematical Biology 48: 279-292.

Coldman, A. J., Goldie, J. H. and Ng, V. 1985. The effect of cellular differentiation on the development of permanent drug resistance. Mathematical Biosciences 74: 177-198.

Cooper, S. 1979. A unifying model for the G1 period in prokaryotes and eukaryotes. Nature 280: 17-19.

Cooper, S. 1984. The continuum model as a unified description of the division cycle of eukaryotes and prokaryotes. In: The Microbial Cell Cycle (Nurse, P. and Streiblova, E., eds.) CRC Press, Boca Raton, FL, pp. 8-27.

Cooper, S. 1991. Bacterial Growth and Division : Biochemistry and Regulation of Prokaryotic and Eukaryotic Division Cycles. Academic Press, San Diego.

Counter, C.M., Avilion, A.A., Lefeuvre, C.E., Stewart, N.G., Greider, C.W., Harley, C.B. and Bacchetti, S. 1992. Telomere shortening associated with chromosome instability is arrested in immortal cells which express telomerase activity. EMBO Journal 11: 1921-1929.

Cowan R. 1985. Branching process results in terms of moments of the generation-time distribution. Biometrics 41: 681-689.

Cowan R. and Culpin D. 1981. A method for the measurement of variability in cell lifetimes. Mathematical Biosciences 54: 249-263.

Cowan R. and Morris V.B. 1986. Cell population dynamics during the differentiation phase of tissue development. Journal of Theoretical Biology 122: 205-224.

Cowan, R. and Staudte, R. 1986. The bifurcating autoregression model in cell lineage studies. Biometrics 42: 769-783.

Crump, K. S. 1970. On systems of renewal equations. Journal of Mathematical Analysis and Applications 30: 425-434.

Crump, K.S. and Hoel, D.G. 1974. Mathematical models for estimating mutation rates in cell populations. Biometrika 61: 237-252.

Crump, K. S. and Mode, C.J. 1969. An age-dependent branching process with correlations among sister cells. Journal of Applied Probability 6: 205-210.

Czerniak, B., Herz, F., Wersto, R.P. and Koss, L.G. 1992. Asymmetric distribution of oncogene products at mitosis. Proceedings of the National Academy of Sciences USA 89: 4860-4863.

Darzynkiewicz, Z., Carter, S. and Kimmel, M. 1984. Effects of [3H]Udr on the cell-cycle progression of L1210 cells. Cell and Tissue Kinetics 17: 641-655.

Darzynkiewicz, Z., Traganos, F. and Kimmel, M. 1986. Assay of cell cycle kinetics by multivariate flow cytometry using the principle of stathmokinesis. In: Techniques in Cell Cycle Analysis (Gray, J.E. and Darzynkiewicz, Z., eds.). Humana Press, Clifton, NJ, pp. 291-336.

Darzynkiewicz, Z., Crissman, H., Traganos, F. and Steinkamp, J. 1982. Cell heterogeneity during the cell cycle. Journal of Cellular Physiology 113: 465-474.

Darzynkiewicz, Z., Evenson, D. P., Staiano-Coico, L., Sharpless, T.K., Melamed, M. L. 1979. Correlation between cell cycle duration and RNA content. Journal of Cellular Physiology 100: 425-438.

Dawson, D.A. and Hochberg, K.J. 1991. A multilevel branching model. Advances in Applied Probabiltiy 23: 701-715.

Dawson, D. and Perkins, E. 1991. Historical processes. Memoirs of the American Mathematical Society 93(454).

Day, R. S. 1986a. Treatment sequencing, asymmetry, and uncertainty: Protocol strategies for combination chemotherapy. Cancer Research 46: 3876-3885.

Day, R. 1986b. A branching process model for heterogeneous cell populations. Mathematical Biosciences 78: 73-90.

Demetrius, L., Schuster, P. and Sigmund, K. 1985. Polynuclotide evolution and branching processes. Bulletin of Mathematical Biology 47: 239-262.

Demos, J.P. 1982. Entertaining Satan: Witchcraft and the Culture of early New England. Oxford University Press, New York.

Dibrov, B.F., Zhabotinsky, A.M., Neyfakh, Y.A., Orlova, M.P. and L.I. Churikova, L.I. 1985. Mathematical model of cancer chemotherapy. Periodic schedules of phase specific cytotoxic agent administration increasing the selectivity of therapy. Mathematical Biosciences 73: 1-31.

Dibrov, B.F., Zhabotinsky, A.M., Neyfakh, Y.A., Orlova, M.P. and Churikova, L.I. 1983. Optimal scheduling for cell synchronization by cycle-specific blockers. Mathematical Biosciences 66: 167-185.

Doetsch, G. 1974. Introduction to the Theory and Application of the Laplace Transformation. Springer-Verlag, New York.

Durbin, R. Eddy, S. Krogh, A. and Mitchison, G. 1998. Biological Sequence Analysis: Probabilistic Models of Proteins and Nucleic Acids." Cambridge University Press, Cambridge.

Durrett, R. 1978. The genealogy of critical branching processes. Stochastic Processes and Their Applications 8: 101-116.

Etheridge, A.M. 1992. Conditioned superprocesses and a semilinear heat equation. In: Seminar on Stochastic Processes (Seattle, WA, 1992) (Cinlar, E., Chung, K.L. and Sharpe, M.J., eds.). Progress in Probability, 33. Birkhäuser, Boston, pp. 89-99.

Falahati, A. 1999. Two-sex branching populations. Doctoral thesis. Department of Mathematical Statistics, Chalmers University, Göteborg, Sweden. Dissertations Series no. 1493.

Fearn, D.H. 1972. Galton–Watson processes with generation dependence. Proceedings of the Sixth Berkley Symposium on Mathematical and Statistical Probabilitity 4: 159-172.

Fearn, D.H. 1976. Supercritical age dependent branching process with generation dependence. The Annals of Probability 4: 27-37.

Feller, W. 1968. An Introduction to Probability and Its Applications. Vol. 1, 3rd ed., Wiley, New York.

Feller, W. 1971. An Introduction to Probability and Its Applications. Vol. 2, 2nd ed., Wiley, New York.

Fleischmann, K. and Siegmund-Schultze, R. 1977. The structure of the reduced critical Galton–Watson processes. Mathematische Nachrichten 79: 233-241.

Fleischmann, K. and Vatutin, V.A. 1999. Reduced subcritical Galton–Watson processes in random environment. Advances in Applied Probability 31: 1-24.

Gawel, B. and Kimmel, M. 1996. Iterated Galton–Watson process. Journal of Applied Probability 33: 949-959.

Gillespie, J.H. 1986. Variability of evolutionary rates of DNA. Genetics 113: 1077-1091.

Goldie, A.J. 1982. Drug resistance and chemotherapeutic strategy. In: Tumor Cell Heterogeneity (Owens, A.H., Coffey, D.S. and Baylin, S.B., eds.). Academic Press, New York, pp. 115-125.

Goldie, J.H. and Coldman, A.J. 1979. A mathematical model for relating the drug sensitivity of tumors to their spontaneous mutation rate. Cancer Treatment Reports 63: 1727-1733.

Goldie, J.H. and Coldman, A.J. 1984. The genetic origin of drug resistance in neoplasms: Implications for systemic therapy. Cancer Research 44: 3643-3653.

Goldie, J.H., Coldman, A.J. and Gudauskas, G.A. 1982. Rationale for the use of alternating non-cross-resistant chemotherapy. Cancer Treatment Reports 66: 439-449.

González, M. and Molina, M. (1996) On the limit behaviour of a superadditive bisexual Galton–Watson branching process. Journal of Applied Probability 33: 960–967.

Greider, C.W. 1996. Telomere length regulation. Annual Review of Biochemistry 65: 337-365.

Greider, C.W. and Blackburn, E.H. 1996. Telomeres, telomerase and cancer. Scientific American 274 (2): 92-97.

Griffiths, R.C. and Tavaré, S. 1999. The ages of mutations in gene trees. Annals of Applied Probability 9: 567–590.

Gusev, Y. and Axelrod, D.E. 1995. Evaluation of models of inheritance of cell cycle times: Computer simulation and recloning experiments. In: Mathematical Population Dynamics: Analysis of Heterogeneity. Vol. 2. Carcinogenesis and Cell & Tumor Growth (Arino, A., Axelrod, D. and Kimmel, M., eds.). Wuerz Publishing, Winnipeg, Ontario, Canada, pp. 97-116.

Guttorp P. 1991. Statistical Inference for Branching Processes. Wiley Series in Probability and Mathematical Statistics. Wiley, New York.

Gyllenberg, M. 1986. The size and scar distributions of the yeast *Saccharomyces cerevisiae*. Journal of Mathematical Biology 24: 81-101.

Harley, C. B. 1991. Telomere loss: Mitotic clock or genetic time bomb? Mutation Research 256: 271-282.

Harley, C.B. and Goldstein, S. 1980. Retesting the commitment theory of cellular aging. Science 207: 191-193.

Harnevo, L.E. and Agur, Z. 1991. The dynamics of gene amplification described as a multitype compartmental model and as a branching process. Mathematical Biosciences 103: 115-138.

Harnevo, L.E. and Agur, Z. 1992. Drug resistance as a dynamic process in a model for multistep gene amplification under various levels of selection stringency. Cancer Chemotherapy and Pharmacology 30: 469-476.

Harnevo, L.E. and Agur, Z. 1993. Use of mathematical models for understanding the dynamics of gene amplification. Mutation Research 292: 17-24.

Harpending, H.C., Batzer, M.A., Gurven, M., Jorde, L.B., Rogers, A.R. and Sherry S.T. 1998. Genetic traces of ancient demography. Proceedings of the National Academy of Sciences USA 95: 1961-1967

Harris, T.E. 1963. The Theory of Branching Processes. Springer-Verlag, Berlin.

Hasegawa, M. and Horai, S. 1990. Time of the deepest root for polymorphism in human mitochondrial DNA. Journal of Molecular Evolution 32: 37-42.

Hästbacka, J., de la Chapelle, A., Kaitila, I., Sistonen, P., Weaver, A. and Lander, E. 1992. Linkage disequilibrium mapping in isolated founder populations: Diastrophic dysplasia in Finland. Nature Genetics 2: 204-211.

Hejblum, G., Costagiola, D., Valleron, A.-J. and Mary, J.-Y. 1988. Cell cycle models and mother–daughter correlation. Journal of Theoretical Biology 131: 255-262.

Hoel, D.G. and Crump K.S. 1974. Estimating the generation-time distribution of an age-dependent branching process. Biometrics 30: 125-135.

International Human Genome Sequencing Consortium. 2001. Initial sequencing and analysis of the human genome. Nature 409: 860-921.

Jagers, P. 1975. Branching Processes with Biological Applications. Wiley, London.

Jagers, P. 1983. Stochastic models for cell kinetics. Bulletin of Mathematical Biology 45: 507-519.

Jagers, P. 1991. The growth and stabilization of populations. Statistical Science 6: 269-283.

Jagers, P. 1992. Stabilities and instabilities in population dynamics. Journal Applied Probability 29: 770-780.

Jagers, P. 1995. Dependence in branching. Preprint 34: 1-17.

Jagers, P. 2001. The deterministic evolution of general branching populations. IMS Lecture Notes and Monographs Series, 36: 384-398.

Jagers, P. and Nerman, O. 1996. The asymptotic composition of supercritical multi-type branching populations (Mar Yor, ed.). Séminaire de Probabilités. Lecture Notes in Mathematics. Springer-Verlag, Berlin, pp. 40-54.

Jagers, P. and Norrby, K. 1974. Estimation of the mean and variance of cycle times in cinemicrographically recorded cell populations during balanced exponential growth. Cell and Tissue Kinetics 7: 201-211.

Joffe, A. and Waugh, W. 1982. Exact distributions of kin numbers in a Galton–Watson process. Journal Applied Probability 19: 767-775.

Joffe, A. and Waugh, W. 1985. The kin number problem in a multitype Galton–Watson population. Journal Applied Probabitlity 22: 37-47.

Joffe, A. and Waugh, W. 1986. Exact distributions of kin numbers in a multitype Galton–Watson process. In: Semi-Markov Models (Janssen, J., ed.). Plenum Press, New York, pp. 397-405.

Jones, M.E. 1994. Luria–Delbrück fluctuation experiments; accounting simultaneously for plating efficiency and differential growth rate. Journal of Theoretical Biology 166: 355-363.

Jones, M.E., Thomas, S.M. and Rogers, A. 1994. Luria–Delbrück fluctuation experiments: Design and analysis. Genetics 136: 1209-1216.

Joyce, G.F. 1992. Directed molecular evolution. Scientific American 267(6) 90-97.

Kaplan, N.L., Hill, W.G. and Weir, B.S. 1995. Likelihood methods for locating disease genes in nonequilibrium populations. American Journal of Human Genetics 56: 18-32.

Kaufman, R.J., Brown, P.C. and Schimke, R.T. 1981. Loss and stabilization of amplified dihydrofolate reductase genes in mouse sarcoma S-180 cell lines. Molecular and Cellular Biology 1: 1084-1093.

Kendal, W.S. and Frost, P. 1988. Pitfalls and practice of Luria–Delbrück fluctuation analysis: A review. Cancer Research 48: 1060-1065.

Kesten, H. 1989. Supercritical branching processes with countably many types and the size of random Cantor sets. In: Probability, Statistics, and Mathematics: Papers in Honor of Samuel Karlin (Anderson, T.W., Athreya, K.B. and Iglehart, D.L., eds.). Academic Press, Boston, pp. 103-121.

Kimmel, M. 1980a. Cellular population dynamics. I: Model construction and reformulation. Mathematical Biosciences 48: 211-224.

Kimmel, M. 1980b. Cellular population dynamics. II: Investigation of solutions. Mathematical Biosciences 48: 225-239.

Kimmel, M. 1980c. Time discrete model of cell population dynamics. Systems Science 6: 343-363.

Kimmel, M. 1982. An equivalence result for integral equations with application to branching processes. Bulletin of Mathematical Biology 44: 1-15.

Kimmel, M. 1983. The point process approach to age- and time-dependent branching processes. Advances in Applied Probability 15: 1-20.

Kimmel, M. 1985. Nonparametric analysis of stathmokinesis. Mathematical Biosciences 74: 111-123.

Kimmel, M. 1987. Metabolic events in the cell cycle of malignant and normal cells. A mathematical modeling approach. In: Cancer Modeling (Thompson, J.R. and Brown, B., eds.). Marcel Dekker, New York, pp. 215-235.

Kimmel, M. 1994. Rapid genome evolution and cancer: A modeling perspective. Applied Mathematics and Computer Science 4: 163-177.

Kimmel, M. 1997. Quasistationarity in a branching model of division-within-division. In: Classical and Modern Branching Processes (Athreya, K.B. and Jagers, P., eds.). IMA Volumes in Mathematics And Its Applications, 84. Springer-Verlag, New York, pp. 157-164.

Kimmel, M. and Arino, O. 1991. Cell cycle kinetics with supramitotic control, two cell types and unequal division: A model of transformed embryonic cells. Mathematical Biosciences 105: 47-79.

Kimmel, M. and Axelrod, D.E. 1990. Mathematical models of gene amplification with applications to cellular drug resistance and tumorigenicity. Genetics 125: 633-644.

Kimmel, M. and Axelrod, D.E. 1991. Unequal cell division, growth regulation and colony size of mammalian cells: A mathematical model and analysis of experimental data. Journal of Theoretical Biology 153: 157-180.

Kimmel, M. and Axelrod, D.E. 1994. Fluctuation test for two-stage mutations: Application to gene amplification. Mutation Research 306: 45-60.

Kimmel, M. and Stivers, D. 1994. A time-continuous branching process model of unstable gene amplification. Bulletin of Mathematical Biology 56: 337-357.

Kimmel, M. and Swierniak, A. 1982. On a certain optimal control problem related to the optimal chemotherapy of leukemia. Technical Reports of the Silesian Technical University, Series Automation (Zeszyty Naukowe Politechniki Slaskiej, Seria Automatyka) 65: 121-130. (in Polish, English abstract).

Kimmel, M. and Traganos, F. 1985. Kinetic analysis of drug induced G2 block in vitro. Cell and Tissue Kinetics 18: 91-110.

Kimmel, M. and Traganos, F. 1986. Estimation and prediction of cell cycle specific effects of anticancer drugs. Mathematical Biosciences 80: 187-208.

Kimmel, M., Axelrod, D.E. and Wahl, G.M. 1992. A branching process model of gene amplification following chromosome breakage. Mutation Research 276: 225-239.

Kimmel, M., Darzynkiewicz, Z. and Staiano-Coico, L.1986. Stathmokinetic analysis of human epidermal cells in vitro. Cell and Tissue Kinetics 19: 289-304.

Kimmel, M., Traganos, F. and Darzynkiewicz, Z. 1983. Do all daughter cells enter the 'Indeterminate' ('A') state of the cell cycle? Analysis of stathmokinetic experiment on L1210 cells. Cytometry 4: 191-201.

Kimmel, M., Darzynkiewicz, Z., Arino, O. and Traganos, F. 1984. Analysis of a cell cycle model based on unequal division of metabolic constituents to daughter cells during cytokinesis. Journal of Theoretical Biology 110: 637-664.

Kimmel, M., Grossi, A., Amuasi, J. and Vannucchi, A.M. 1990. Non-parametric analysis of platelet lifespan. Cell and Tissue Kinetics 23: 191-202.

Kimmel M., Chakraborty, R., King, J.P., Bamshad, M., Watkins, W.S. and Jorde, L.B. 1998. Signatures of population expansion in microsatellite repeat data. Genetics 148: 1921-1930.

Kirkwood, T.B.L. and Holliday, R. 1978. A stochastic model for the commitment of human cells to senescence. In: Biomathematics and Cell Kinetics (Valleron, A.J. and Macdonald, P.D., eds.). Elsevier/North-Holland, Amsterdam, pp. 161-172.

Klebaner, F. 1988. Conditions for fixation of an allele in the density-dependent Wright–Fisher models. Journal of Applied Probability 25: 247-256.

Klebaner, F. 1990. Conditions for the unlimited growth in multitype population size dependent Galton–Watson processes. Bulletin of Mathematical Biology 52: 527-534.

Klebaner, F. 1997. Population and density dependent branching process. In Classical and Modern Branching Processes (Athreya, K.B. and Jagers, P., eds.). IMA Volumes in Mathematics and Its Applications, 84. Springer, New York pp. 165-170.

Klebaner, F. and Zeitouni, O. 1994. The exit problem for a class of density dependent branching systems. Annals of Applied Probability 4: 1188-1305.

Klein, B. and Macdonald, P.D.M. 1980. The multitype continuous-time Markov branching process in a periodic environment. Mathematical Sciences 12: 1-13.

Knolle, H. 1988. Cell Kinetic Modelling and the Chemotherapy of Cancer. Springer-Verlag, Berlin.

Koteeswaran, P. 1989. Estimating the age of a Galton–Watson process with binomial offspring distribution. Stochastic Analysis and Applications 7: 413–423.

Kotenko, J.L. Miller, J.H. and Robinson, A.I. 1987. The role of asymmetric cell division in *Pteripodhyte* cell differentiation. I. Localized metal accumulation and differentiation in *Vittaria gemmae* and *Onoclea prothallia.* Protoplasma 136: 81-95.

Kowald, A. 1997. Possible mechanisms for the regulation of telomere length. Journal of Molecular Biology 273: 814-825.

Kuczek, T. 1984. Stochastic modeling for the bacterial cell cycle. Mathematical Biosciences 69: 159-171.

Kuczek, T. and Axelrod, D.E. 1986. The importance of clonal heterogeneity and interexperiment variablity in modeling the eukaryotic cell cycle. Mathematical Biosciences 79: 87-96.

Kuczek, T. and Axelrod, D.E. 1987. Tumor cell heterogeneity: Divided-colony assay for measuring drug response. Proceedings of the National Academy Sciences USA 84:4490-4494.

Kuczek, T. and Chan, T. C. K. 1988. Mathematical modeling for tumor resistance. Journal of the National Cancer Institute 80: 146-147. (Response: Goldie, J. H. and Coldman, A. J. 1988. Journal of the National Cancer Institute 80: 146-147.)

Lapidus, R. 1984. Growth and division kinetics of asymmetrically dividing *Tetrahymena thermophilia*. Journal of Theoretical Biology 106: 135-140.

Larson, D.D., Spangler, E.A. and Blackburn, E.H. 1987. Dynamics of telomere length variation in *Tetrahymena thermophila*. Cell 50: 477-483.

Lea, D.E. and C.A. Coulson, C.A. 1949. The distribution of the numbers of mutants in bacterial populations. Journal of Genetics 49: 264-265.

Levy, S.B. 1998. The challenge of antibiotic resistance. Scientific American 278(3): 46-53.

Levy, M.Z., Allsopp, R.C., Futcher, A.B., Greider, C.W. and Harley, C.B. 1992. Teleomere end-replication problem and cell aging. Journal of Molecular Biology 225: 951-960.

Lewontin, R.C. 2000. The Triple Helix: Gene, Organism, and Environment. Harvard University Press, Cambridge, MA.

Lipow, C.1975. A branching model with population size dependence. Advances in Applied Probability 7: 495-510.

Loeffler, M. and Wichmann, H.E. 1980. A comprehensive mathematical model of stem cell proliferation which reproduces most of the published experimental results. Cell and Tissue Kinetics 13: 543-561.

Luria, S.E. and Delbrück, M. 1943. Mutations of bacteria from virus sensitivity to virus resistance. Genetics 28: 491-511.

Ma, W.T., Sandri, G. vH. and Sarkar, S. 1992. Analysis of the Luria–Delbrück distribution using discrete convolution powers. Journal of Applied Probability 29: 255-267.

Macdonald, P.D.M. 1978. Age distributions in the general cell kinetic model. In: Biomathematics and Cell Kinetics (Valleron, A.J. and Macdonald, P.D.M., eds.). Elsevier/North-Holland Biomedical Press, Amsterdam, pp. 3-20.

Macken, C.A. and Perelson, A.S. 1985. Branching Processes Applied to Cell Surface Aggregaton Phenomena. Springer-Verlag, Berlin.

Macken, C.A. and Perelson, A.S. 1988. Stem Cell Proliferation and Differentiation. A Multitype Branching Process Model. Lecture Notes in Biomathematics, 76. Springer-Verlag, Berlin.

Mackillop, W.J. 1986. Instrinsic versus acquired drug resistance. Cancer Treatment Reports 70: 817. (Reply: Goldie, J.H. and Coldman, A.J. 1986. Cancer Treatment Reports 70: 818.)

Maddox, J. 1992. Is molecular biology yet a science? Nature 355: 201.

Metz, J.A.J. and Diekmann, O. (eds.). 1986. The Dynamics of Physiologically Structured Populations. Lecture Notes in Biomathematics, 68. Springer-Verlag, Berlin.

Mode, C.J. 1971. Multitype Branching Processes. Elsevier, New York.

Morris, V.B. and Cowan, R. 1984. A growth curve of cell numbers in the neural retina of embryonic chicks. Cell and Tissue Kinetics 17: 199-208.

Morris, V.B. and Taylor, I.W. 1985. Estimation of nonproliferating cells in the neural retina of embryonic chicks by flow cytometry. Cytometry 6: 375-380.

Morrow, J. 1970. Genetic analysis of azaguanine resistance in an established mouse cell line. Genetics 65: 279-287.

Moy, S.C. 1967. Extensions of a limit theorem of Everett, Ulam and Harris on multitype branching processes to a branching process with countably many types. Annals of Mathematical Statistics 38: 992–999.

Murnane, J.P. and Yezzi, M.J. 1988. Association of high rate of recombination with amplification of dominant selectable gene in human cells. Somatic Cell and Molecular Genetics 14: 273-286.

Nagylaki, T. 1990. Models and approximations for random genetic drift. Theoretical Population Biology 37: 192–212.

Navidi, W., Tavare, S. and Arnheim, N. 1996. The role of the mutation rate and selective pressures on observed levels of the human mitochondrial DNA deletion mtDNA 4977. Unpublished manuscript.

Nedelman, J., Downs, H. and Pharr, P. 1987. Inference for an age-dependent, multitype branching-process model of mast cells. Journal of Mathematical Biology 25: 203-226. (Erratum: Journal of Mathematical Biology 25: 571).

Neveu, J. 1975. Discrete-Parameter Martingales. rev. ed. Speed, T.P., North-Holland Mathematical Library, 10. North-Holland (Amsterdam/transl.). American Elsevier, New York.

O'Connell, N. 1993. Yule process approximation for the skeleton of a branching process. Journal Applied Probability 30: 725-729.

O'Connell, N. 1995. The genealogy of branching processes and the age of our most recent common ancestor. Advances in Applied Probability 27: 418-442.

Olofsson, P. 1996. Branching processes with local dependencies. The Annals of Applied Probability 6: 238-268.

Olofsson, P. 2000. A branching process model of telomere shortening. Communications in Statistics. Stochastic Models 16: 167–177.

Olofsson, P. and Kimmel, M. 1999. Stochastic models of telomere shortening. Mathematical Biosciences 158: 75–92.

Olofsson, P. and Shaw, C. 2001. Exact sampling formulas for multi-type Galton–Watson processes. Journal of Mathematical Biology, to appear.

Olofsson, P., Schwalb, O., Chakraborty, R., and Kimmel, M. 2001. An application of a general branching process in the study of the genetics of aging. Journal of Theoretical Biology 213: 547–557.

Olovnikov, A.M. 1973. A theory of marginotomy. Journal of Theoretical Biology 41: 181-190.

Pakes, A.G. 1993. Explosive Markov branching processes: Entrance laws and limiting behaviour. Advances in Applied Probability 25: 737–756.

Pakes, A.G. 1994. On the recognition & structure of probability generating functions. Research Report, Department of Mathematics, The University of Western Australia, Nedlands, WA, Australia. pp. 1-29.

Pakes, A.G. 2000. Biological applications of branching processes. Research Report, Department of Mathematics and Statistics, The University of Western Australia, Nedlands, WA, Australia.

Pakes, A.G. and Dekking, F.M. 1991. On family trees and subtrees of simple branching processes. Journal of Theoretical. Probability 4: 353–369.

Pakes, A.G. and Trajstman, A.C. 1985. Some properties of continuous-state branching processes, with applications to Bartoszynski's virus model. Advances in Applied Probability 17: 23-41.

Pankratz, V.S. 1998. Stochastic Models and Linkage Disequilibrium. Doctoral thesis. Department of Statistics, Rice University, Houston, TX.

Peterson, J.A. 1984. Analysis of variability in albumin content of sister hepatoma cells and model for geometric phenotypic variability (Quantitative Shift Model). Somatic Cell and Molecular Genetics 10: 345-357.

Pharr, P.N., Nedelman, J., Downs, H.P., Ogawa, M. and Gross, A.J. 1985. A stochastic model for mast cell proliferation in culture. Journal of Cellular Physiology 125: 379-386.

Polanski, A., Kimmel, M. and Swierniak A. 1997. Qualitative analysis of the infinite model of drug resistance evolution. In: Advances in Mathematical Population Dynamics – Molecules, Cells and Man (Arino, O., Axelrod, D. and Kimmel, M., eds.). World Scientific, Singapore, pp. 595-612.

Polanski, A., Swierniak, A. and Duda, Z. 1993. Multiple solutions to the TPBVP arising in optimal scheduling of cancer chemotherapy. Conference Proceedings 1993, IEEE International Conference on Systems, Man and Cybernetics, Vol. 4, pp. 5-8.

Puck, T.T. and Steffen, J. 1963. Life cycle analysis of mammalian cells. Part I. Biophysical Journal 3: 379-397.

Richards, R.I. and Sutherland, G.R. 1994. Simple repeat DNA is not replicated simply. Nature Genetics 6: 114-116.

Rigney, D.R. 1981. Correlation between the ages of sibling cell cycle events and a test of the "transition probabability" cell cycle model. In: Biomathematics and Cell Kinetics (Rotenberg, M., ed.). Elsevier/North Holland Biomedical Press, Amsterdam. pp. 157-166.

Rittgen, W. 1983. Controlled branching processes and their applications to normal and malignant haematopoiesis. Bulletin of Mathematical Biology 45: 617-626.

Rosen, R. 1986. Role of mathematical modeling in protocol formulation in cancer chemotherapy. Cancer Treatment Reports 40: 1461-1462. (Reply: Coldman, A.J. and Goldie, J.H. 1986. Cancer Treatment Reports 70: 1461-1462).

Sagitov, S. 1989. The limit behavior of reduced critical branching processes. Soviet Mathematics Doklady 38: 488-491.

Saiki, R.K., Gelfand, D.H., Stoffel, S., Scharf, S.J., Higuchi, R., Horn, G.T., Mullis, K.B. and Erlich, H.A. 1988. Primer-directed enzymatic amplification of DNA with a thermostable DNA polymerase. Science 239: 487-491.

Sawyer, S. 1976. Branching diffusion processes in population genetics. Advances in Applied Probability 8: 659-689.

Seneta, E. and Tavaré, S. 1983. Some stochastic models for plasmid copy number. Theoretical Population Biology 23: 241-256.

Sennerstam R. 1988. Partition of protein (mass) to sister cell pairs at mitosis: A re-evaluation. Journal of Cell Science 90: 301-306.

Sennerstam, R. and Strömberg, J.-O. 1984. A comparative study of the cell cycles of nullipotent and mulitpotent embryonal carcinoma cell lines during exponential growth. Developmental Biology 103: 221-229.

Sennerstam R. and Strömberg, J.-O. 1988. Evidence for an intraclonal random shift between two types of cell cycle times in an embryonal carcinoma cell line. Journal of Theoretical Biology 131: 151-162.

Sennerstam, R. and Strömberg, J.-O. 1995. Cell cycle progression: Computer simulation of uncoupled subcycles of DNA replication and cell growth. Journal of Theoretical Biology 175: 177-189.

Sennerstam, R. and Strömberg, J.-O. 1996. Exponential growth, random transitions and progress through the G1 phase: Computer simulation of experimental data. Cell Proliferation 29: 609-622.

Shaw, C.A. 2000. Genealogical methods for multitype branching processes with applications in biology. Ph.D dissertation, Department of Statistics, Rice University, Houston, TX.

Shenkar, R., Navidi, W., Tavare, S., Dang, M. H., Chomyn A., Attardi, G., Cortopassi, G., and Arnheim, N. 1996. The mutation rate of the human mtDNA deletion mtDNA4977. American Journal of Human Genetics 59:772-780

Spătaru, A. 1989. Properties of branching processes with denumerably many types. Revue Roumaine de Mathématiques Pures et Appliquées (Romanian Journal of Pure and Applied Mathematics) 34: 747-759.

Staiano-Coico, L., Hajjar, D.P., Hefton, J.M., Hajjar, K. and Kimmel, M. 1988. Interaction of arterial cells: III. Stathmokinetic analyses of smooth muscle cells cocultured with endothelial cells. Journal of Cellular Physiology 134: 485-490.

Staudte, R.G. 1992. A bifurcating autoregression model for cell lineage data with varying generation means. Journal of Theoretical Biology 156: 183-195.

Staudte, R.G., Guiguet, M. and d'Hooghe, M.C. 1984. Additive models for dependent cell populations. Journal of Theoretical Biology 109: 127-146.

Staudte, R.G., Huggins, R.M., Zhang, J., Axelrod, D.E. and Kimmel, M. 1997. Estimating clonal heterogeneity and interexperiment variability with the bifurcating autoregression model for cell lineage data. Mathematical Biosciences 143: 103-121.

Stewart, F.M., Gordon, D.M. and Levin, B.R. 1990. Fluctuation analysis: The probability distribution of the number of mutants under different conditions. Genetics 124: 175-185.

Stigler, S.M. 1970. Estimating the age of a Galton–Watson branching process. Biometrika 57: 505-512.

Stivers, D.N. and Kimmel, M. 1996a. A continuous-time, multi-type generational inheritance branching process model of cell proliferation with clonal memory. Nonlinear World 3: 385-399.

Stivers, D.N. and Kimmel, M. 1996b. On the clonal inheritance model of cell proliferation. Proceedings of the First World Congress of Nonlinear Analysts, Tampa, Florida, August 1992. Walter de Gruyter, Berlin, pp. 3401-3408.

Stivers, D.N., Kimmel, M. and Axelrod, D.E. 1996. A discrete-time, multi-type generational inheritance branching process model of cell proliferation. Mathematical Biosciences 137: 25-50.

Stoneking, M., Sherry, S.T., Redd, A.J. and Vigilant, L. 1992. New approaches to dating suggest a recent age for the human mtDNA ancestor. Philosophical Transactions of the Royal Society of London B, Biological Sciences 337: 167-175.

Swierniak A. and Kimmel, M. 1984. Optimal control application to leukemia chemotherapy protocols design. Technical reports of the Silesian Technical University, Series Automation (Zeszyty Naukowe Politechniki Slaskiej, Seria Automatyka) 73: 261-277(in Polish, English abstract).

Swierniak A. and Kimmel, M. 1991. Cancer cell synchronization and recruitment as optimal control problems. Proceedings of 13th World IMACS Congress, Dublin, Vol. 3, pp. 1461-1462.

Swierniak, A. Polanski, A. and Kimmel, M. 1996. Control problems arising in chemotherapy under evolving drug resistance. Preprints of the 13th World Congress of IFAC, Volume B, 411-416.

Taïb, Z. 1987. Labelled branching processes with applications to neutral evolution theory. Ph.D. thesis, Chalmers University of Technology, Sweden.

Taïb, Z. 1992. Branching Processes and Neutral Evolution. Lecture Notes in Biomathematics, 93. Springer-Verlag, Berlin.

Taïb, Z. 1993. A note on modeling the dynamics of budding yeast populations using branching process. Journal of Mathematical Biology 31: 805-815.

Taïb, Z. 1995. Branching processes and functional-differential equations determining steady-size distributions in cell populations. Journal of Applied Probability 32: 1-10.

Tan, W.Y. 1982. On the distribution theories for the number of mutants in cell populations. SIAM Journal of Applied Mathematics 42: 719-730.

Tan, W.Y. 1983. On the distribution of the number of mutants at the hypoxanthine–quanine phosphoribosyl transferase locus in Chinese hamster ovary cells. Mathematical Biosciences 67: 175-192.

Tannock, I. 1978. Cell kinetics and chemotherapy: A critical review. Cancer Treatment Reports 62: 1117-1133.

Tavaré, S. 1980. Time-reversal and age distribution. 1. Discrete Markov case. Journal of Applied Probability 17: 33-46.

Tavaré, S. 1984. Line-of-descent and genealogical processes, and their applications in population genetics models. Theoretical Population Biology 26: 119-164.

Therneau, T.M., Solberg, L.A. Jr. and Jenkins, R.B. 1989. Modeling megakaryocyte development as a branching process. Computers and Mathematics with Applications 18: 959-964.

Till, J.E., McCulloch, E.A. and Siminovitch, L. 1964. A stochastic model of stem cell proliferation, based on the growth of spleen colony-forming cells. Proceedings of the National Academy of Sciences USA 51: 29-36.

Tltsy, T., Margolin, B.H. and Lum, K. 1989. Differences in the rates of gene amplification in nontumorigenic and tumorigenic cell lines as measured by Luria-Delbrück fluctuation analysis. Proceedings of the National Academy of Sciences USA 86: 9441-9445.

Tyrcha, J. 1988. Asymptotic stability in a generalized probabilistic/deterministic model of the cell cycle. Journal of Mathematical Biology 26: 465-475.

Tyson, J.J. 1987. Size control of cell division. Journal of Theoretical Biology 126: 381-391.

Tyson, J.J. and Hannsgen, K.B. 1985a. Global asymptotic stability of the size distribution in probabilistic models of the cell cycle. Journal of Mathematical Biology 22: 61-68.

Tyson, J.J. and Hannsgen, K.B. 1985b. The distributions of cell size and generation time in a model of the cell cycle incorporating size control and random transitions. Journal of Theoretical Biology 113: 29-62.

Tyson, J.J. and Hannsgen, K.B. 1986. Cell growth and division: A deterministic/probabilistic model of the cell cycle. Journal of Mathematical Biology 23: 231-246.

Tyson, J., Garcia-Herdugo, G. and Sachsenmaier, W. 1979. Control of nuclear division in *Physarum polycephalum*. Experimental Cell Research 119: 87-98.

Varshaver, N.B., Marshak, M.I. and Shapiro, N.I. 1983. The mutational origin of serum independence in Chinese hamster cells in vitro. International Journal of Cancer 31: 471-475.

Venter, J.C. et al. 2001. The sequence of the human genome. Science 291: 1304-1351.

Vigilant, L.R., Pennington, H., Harpending, H., Kocher, T.D. and Wilson, A. 1989. Mitochondrial DNA sequences in single hairs from a southern African population. Proceedings of the National Academy of Sciences USA 86: 9350-9354.

Vigilant, L., Stoneking, R., Harpending, H., Hawkes, K. and Wilson, A. 1991. African populations and the evolution of human mitochondrial DNA. Science 253: 1503-1507.

Vogel, H., Niewisch, H. and Matioli, G. 1969. Stochastic development of stem cells. Journal Theoretical Biology 22: 249-270.

Waugh, W.A.O'N. 1981. Application of the Galton–Watson process to the kin number problem. Advances in Applied Probability 13: 631-649.

Webb, G.F. 1987. Random transitions, size control, and inheritance in cell population dynamics. Mathematical Biosciences 85: 71-91.

Webb, G.F. 1989. Alpha- and beta-curves, sister–sister and mother–daughter correlations in cell population dynamics. Computers and Mathematics with Applications 18: 973-984.

Weiss, G. and von Haeseler, A. 1997. A coalescent approach to the polymerase chain reaction. Nucleic Acids Research 25: 3082-3087.

Wilson, A.C. and Cann, R.L. 1992. Recent African genesis of humans. Scientific American 266(4): 68-73.

Windle, B., Draper, B.W., Yin, Y., O'Gorman, S. and Wahl, G.M. 1991. A central role for chromosome breakage in gene amplification, deletion formation, and amplicon integration. Genes & Development 5: 160-174.

Yakovlev, A.Yu. and Yanev, N.M. 1989. Transient Processes in Cell Proliferation Kinetics. Springer-Verlag, Berlin.

Yule, U.G. 1924. A mathematical theory of evolution based on conclusions of Dr. J.C. Willis, F.R.S. Philosophical Transactions of the Royal Society of London, Series B 213: 21-87.

Zakian, V.A. 1995. Telomeres: Beginning to understand the end. Science 270: 1601-1607.

Zakian, V.A. 1996. Structure, function, and replication of *Saccharomyces cerevisiae* telomeres. Annual Review of Genetics 30: 141-172.

Zubkov, A.M. 1975. Limiting distribution for the distance to the closest mutual ancestor. Theory of Probability and Its Applications 20: 602-612.

APPENDIX A

Multivariate Probability Generating Functions

In this section, we will collect some results which are referred to throughout the book. Suppose $X = (X_1, \ldots, X_n) \sim \{p_{i_1 i_2 \ldots i_n}\}_{i_1, i_2, \ldots, i_n \geq 0}$ is a finite vector of non-negative random variables, or a Z_+^n-valued rv.

Definition 6 (Definition of the multivariate pgf). The pgf f_X of a Z_+^n-valued rv X is the function

$$f_X(s) = \mathrm{E}\left(s_1^{X_1} s_2^{X_2} \cdots s_n^{X_n}\right) = \sum_{i_1, i_2, \ldots, i_n \geq 0} p_{i_1 i_2 \cdots i_n} s_1^{i_1} s_2^{i_2} \cdots s_n^{i_n}, \tag{A.1}$$

well defined if $s = (s_1, s_2, \ldots, s_n) \in U_n \equiv [0, 1]^n$.

Theorem 30 (Multivariate pgf theorem). *Suppose X is a Z_+^n-valued rv with pgf f_X. Let us denote by (N_i) the nontriviality condition for the ith coordinate of X, that is, $P[X_i \leq 1] < 1$.*

1. *f_X is non-negative and continuous with all derivatives. Under (N_i), it is increasing and convex as a function of s_i.*
2. *The marginal laws for subsets of X_i's can be obtained by setting respective arguments of the pgf equal to 1 [e.g., $f_X(s)_{|s_j=1,\ j \neq i} = f_{X_i}(s_i)$, etc.]; $f_X(e) = 1$, where $e = (1, \ldots, 1)$.*
3. $\partial^{k_1 + \cdots + k_n} f_X(0) / \partial s_1^{k_1} \cdots \partial s_n^{k_n} = k_1! \cdots k_n!\, p_{k_1 \cdots k_n}$.
4. *The $(k_1, \ldots, k_n)$th mixed factorial moment of X,* $\mu_{k_1, \ldots, k_n} = \mathrm{E}\left[\prod_{i=1}^n \prod_{j=0}^{k_i - 1} (X_i - j)\right]$, *is finite if and only if* $\partial^{k_1 + \cdots + k_n} f_X(e-) / \partial s_1^{k_1} \cdots \partial s_n^{k_n} = \lim_{s \uparrow e} \partial^{k_1 + \cdots + k_n} f_X(s) / \partial s_1^{k_1} \cdots \partial s_n^{k_n}$ *is finite. In such a case,* $\mu_{k_1, \ldots, k_n} = \partial^{k_1 + \cdots + k_n} f_X(e-) / \partial s_1^{k_1} \cdots \partial s_n^{k_n}$.
5. *If X and Y are two independent Z_+^n-valued rv's, then $f_{X+Y}(s) = f_X(s) f_Y(s)$.*

6. *If Y is a Z_+^n-valued rv and $\left\{X_j^{(i)};\ i \geq 1\right\}$, $j = 1, \ldots, n$, are sequences of Z_+^m-valued rv's, then $V = \sum_{j=1}^{n} \sum_{i_j=1}^{Y_j} X_j^{(i_j)}$ is a Z_+^m-valued rv with pgf $f_V(s) = f_Y\left[f_{X_1^{(1)}}(s), \ldots, f_{X_n^{(1)}}(s)\right]$, $s \in U_m$.*
7. *Suppose $\{X_i;\ i \geq 1\}$ is a sequence of Z_+^n-valued rv's. The limit $\lim_{i\to\infty} f_{X_i}(s) = f_X(s)$ exists for each $s \in U^n$ if and only if the sequence $\{X_i;\ i \geq 1\}$ converges in distribution (i.e., when $\lim_{i\to\infty} \mathrm{P}[X_{i,1} = k_1, \ldots, X_{i,n} = k_n] = \mathrm{P}[X_1 = k_1, \ldots, X_n = k_n]$). Then, $f_X(s)$ is the pgf of the limit rv X.*

A further generalization to the denumerable infinite case is possible. Suppose that $X = (X_1, \ldots, X_n, \ldots) \sim \left\{\{p_{i_1 i_2 \cdots i_n}\}_{i_1, i_2, \ldots, i_n \geq 0}\right\}_{n\geq 1}$ is an infinite vector of non-negative random variables, with the σ-algebra generated by the finite-dimensional truncations of the sequence. Also, we may consider X a Z_+^∞-valued rv.

Definition 7 (Denumerable pgf definition). The pgf f_X of a Z_+^∞-valued rv X is a function

$$f_X(s) = \mathrm{E}\left(s_1^{X_1} s_2^{X_2} \cdots s_n^{X_n} \cdots\right) = \sum_{i_1, i_2, \ldots, i_n \geq 0} p_{i_1 i_2 \cdots i_n} s_1^{i_1} s_2^{i_2} \cdots s_n^{i_n} \tag{A.2}$$

defined for

$$s \in \bigcup_{n\geq 1} U_n \equiv \bigcup_{n\geq 1} \{(s_1, s_2, \ldots, s_n, 1, 1, \ldots) : s_1, s_2, \ldots, s_n \in [0, 1]\} \tag{A.3}$$

(i.e., for arguments $s \in [0, 1]^\infty$ with only finite number of coordinates not equal to 1).

Properties 1 through 5 stated in the multivariate pgf theorem carry over to the finite-dimensional restrictions of the denumerable pgf. An important difference is that Property 6 does not necessarily hold for infinite n, as the resulting sum may be improper (if it is proper, then Property 6 holds). Also, the convergence property (Property 7) requires an additional continuity requirement:

Denumerable pgf convergence. Suppose $\{X_i, i \geq 1\}$ is a sequence of Z_+^∞-valued rv's. A necessary and sufficient condition for convergence in distribution of this sequence to a Z_+^∞-valued rv X is that $\lim_{i\to\infty} f_{X_i}(s) = f_X(s)$ exists for each $s \in \bigcup_{n\geq 1} U_n$, and that f_X is pointwise continuous for all sequences $\{s^{(i)}, i \geq 1\}$ with $s^{(i)} \in U_n$. Then, $f_X(s)$ is the pgf of the limit rv X.

APPENDIX B

Probability Distributions for the Bellman–Harris Process

B.1 Construction

We start with a rigorous construction of the probability space of the process, following Chapter 6 of Harris (1963). The elements of the probability space are family histories of the particles.

B.1.1 The families

Let $\mathcal{I}$ be the collection of elements ι, where ι is either 0 or a finite sequence of positive integers $i_1, i_2, \ldots, i_k$. The collection $\mathcal{I}$ is denumerably infinite. The elements ι are enumerated in a sequence $\iota_1, \iota_2, \ldots,$ starting, for example, with $0, 1, 2, 11, 3, 21, 12, 111, \ldots$. The ancestor or founder is denoted by $< 0 >$, whereas $< i_1, i_2, \ldots, i_k >$ denotes the i_kth child of the i_{k-1}th child of $\ldots$, of the i_2th child of the i_1th child of the ancestor.

The family history ω is the sequence $\omega = (l, \nu; l_1, \nu_1; l_2, \nu_2; l_{11}, \nu_{11}; \ldots)$, where l_ι is a non-negative real and represents the length of life of ι, and ν_ι is a non-negative integer and represents the number of children of ι. The collection of all family histories is denoted by Ω. Family history is a redundant description of the particles pedigree in the sense that it enumerates even "nonexistent" children; for example, if $\nu_{ij} = 5$ (the jth child of the ith child of the ancestor has five children), then none of the pairs l_{ijk}, ν_{ijk} for $k > 5$ corresponds to any members of the pedigree.

For each $\omega \in \Omega$, we define a sequence $I_0(\omega), I_1(\omega), \ldots,$ where I_k is a collection of objects $< \iota >$ called the kth generation. The 0th generation $I_0(\omega)$ is the ancestor $< 0 >$ and $I_1(\omega)$ is the set of all objects $< i >$ with $1 \leq i \leq \nu(\omega)$. The succeeding generations are defined inductively: $I_k(\omega)$ is the set of all objects $< i_1 i_2 \ldots i_k >$ such that $< i_1 i_2 \ldots i_{k-1} >$ belongs to $I_{k-1}(\omega)$ and $i_k \leq \nu_{i_1 i_2 \ldots i_{k-1}}(\omega)$. The set of objects $\bigcup_{k=0}^{\infty} I_k(\omega)$ is called the family $I(\omega)$. In view of remarks in the preceding paragraph, more than one family history ω may, in general, correspond to the same family $I(\omega)$.

B.1.2 The number of objects at given time

If the object $< \iota >=< i_1 \cdots i_k >$ belongs to the family $I(\omega)$, it is born at the time $t' = l + l_{i_1} + \cdots + l_{i_1 i_2 \cdots i_{k-1}}$ and dies at the time $t'' = t' + l_{i_1 i_2 \cdots i_k}$; if $t \in [t', t'')$, then the age of the object at t is $t - t'$. Thus, if at time t we count the objects that are alive and have ages $\leq y$, then $< \iota >$ is counted if and only if the following conditions hold (with obvious modifications if $\iota = 0$)

$$\begin{aligned} i_1 \leq \nu, i_2 \leq \nu_{i_1}, \ldots, i_k \leq \nu_{i_1 i_2 \ldots i_{k-1}}, \\ t - y \leq l + l_{i_1} + \cdots + l_{i_1 i_2 \ldots i_{k-1}} \leq t, \\ l + l_{i_1} + \cdots + l_{i_1 i_2 \ldots i_{k-1}} + l_{i_1 i_2 \ldots i_k} > t. \end{aligned} \tag{B.1}$$

The first line in conditions (B.1) means that $< \iota >$ belongs to the kth generation; the second line says that $< \iota >$ was born between $t - y$ and t; the third line says that $< \iota >$ dies after time t.

For each object, ι, let us define $Z_\iota(y, t, \omega)$ to be 1 if conditions (B.1) hold and to be 0 otherwise. Define

$$Z(y, t, \omega) = \sum_{\iota \in \mathcal{I}} Z_\iota(y, t, \omega)$$

and

$$Z(t, \omega) = Z(\infty, t, \omega) = \sum_{\iota \in \mathcal{I}} Z_\iota(\infty, t, \omega).$$

Thus, $Z_\iota(y, t, \omega)$ is 1 if $< \iota >$ is alive and of age $\leq y$ at t and 0 otherwise; $Z(y, t, \omega)$ is the total number of objects of age $\leq y$ at t; $Z(t, \omega)$ is the total number of objects at t. The possibility $Z(y, t, \omega) = \infty$ for some values of y, t, ω is admitted.

Let us note that if $Z(t_0, \omega_0) = 0$ for some t_0 and ω_0, then $Z(t, \omega_0) = 0$ for all $t > t_0$.

B.1.3 Probability measure

Definition 8. The probability measure P is built on the space Ω of family histories ω in the following way.

1. The random variables l_ι are iid with distribution

   $$\mathrm{P}\{l_\iota \leq t\} = G(t),$$

   where G is a right-continuous probability distribution function for which $G(0+) = 0$.

2. The ν_ι's are independent of each other and of the l's, and iid with the pgf

$$f(s) = \sum_{r=0}^{\infty} p_r s^r = \sum_{r=0}^{\infty} \mathrm{P}\{\nu_\iota = r\} s^r,$$

with the trivial cases excluded and $m \equiv f'(1-) < \infty$.

We denote the kth convolution of G with itself by G^{*k} ($G^{*1} = G$). Thus

$$G^{*k}(t) = \int_{0-}^{t+} G^{*(k-1)}(t-y)\, d_y G(y).$$

Because ω corresponds to a denumerable family of independent real-valued random variables, the basic theorem of Kolmogorov ensures that the above assumptions determine uniquely a countably additive probability measure P on the σ -algebra generated by the cylinder sets in Ω. From the definition of $Z(t, \omega)$, it is seen that Z is measurable in (t, ω), where the measurable (t, ω) sets are those generated by rectangles $A \times B$, A being a Borel t-set and B a measurable set in Ω. This conclusion is equivalent to a statement that the family of rv's $\{Z(t, \omega), t \geq 0\}$ is a stochastic process.

B.1.4 The embedded Galton–Watson process and extinction probability

Let $\zeta_k = \zeta_k(\omega)$ be the number of objects in the kth generation I_k, $k = 0, 1, \ldots$. It can be verified that the sequence of random variables $\{\zeta_k, k = 0, 1, \ldots\}$ is a Galton–Watson branching process with generating function $f(s)$ (usually called the embedded Galton–Watson process). The essence of the proof is to verify the property

$$\mathrm{E}(s^{\zeta_{k+1}} | \zeta_1, \zeta_2, \ldots, \zeta_k) = [f(s)]^{\zeta_k}, \tag{B.2}$$

which characterizes the Galton–Watson process. Equation (B.2) is a version of the forward pgf equation (3.5), conditional on ζ_k.

The embedded Galton–Watson process is helpful in proving that the probability of extinction for the Bellman–Harris process is subject to the same rules which govern the Markov versions. Let us note, for example, that if the embedded process becomes extinct for some ω, then the time-continuous process does too, as there is only a finite number of nonvoid generations $I_k(\omega)$ which may last for only a finite time. Thus, $\lim_{k\to\infty} \zeta_k(\omega) = 0$ implies $\lim_{t\to\infty} Z(t, \omega) = 0$. The opposite is, in general, not true. An example can be a situation when all the objects in the kth generation have life lengths $\leq 2^{-k}$ and, consequently, $Z(t) = 0,\ t > 2$. The following result demonstrates that such occurrences have probability 0.

Theorem 31. *Let A be the event $\{\zeta_n > 0$, for each $n\}$ and let B be the event $\{Z(t) > 0$, for each $t \geq 0\}$. If $\mathrm{P}\{A\} > 0$, then $\mathrm{P}\{B|A\} = 1$.*

Corollary 4. *The probability of extinction [i.e., of the event $\bar{B} \equiv \{Z(t) = 0$, for all sufficiently large $t\}$], is equal to the probability of the event $\bar{A}$ [i.e., to the smallest non-negative root q of the equation $s = f(s)$].*

B.2 Integral Equation

B.2.1 Decomposition into subfamilies

If the initial object dies at or before time t, then the objects present at t are its children or their descendants. For a family history $\omega = (l, \nu; l_1, \nu_1; l_2, \nu_2; l_{11}, \nu_{11}; \ldots)$ and each $i = 1, 2, \ldots$, let us define $\omega_i = (l_i, \nu_i; l_{i1}, \nu_{i1}; l_{i2}, \nu_{i2}; l_{i11}, \nu_{i11}; \ldots)$. The ω_i may be interpreted as the family history of $< i >$ and its descendants, although if $\nu < i$, then this family is not actually realized.

For the family history ω_i, let us define the random variables $Z_\iota(y, t, \omega_i)$, $Z(y, t, \omega_i)$, and $Z(t, \omega_i)$ in a way analogous to that in which, for ω, the rv's $Z_\iota(y, t, \omega)$, $Z(y, t, \omega)$, and $Z(t, \omega)$ were previously defined. Suppose that $l(\omega) \in [0, t]$ and $\nu(\omega) > 0$. It can be formally shown using the above definitions that

$$Z(t, \omega) = \sum_{i=1}^{\nu} Z(t - l, \omega_i). \tag{B.3}$$

In view of the fact that

$$I(\omega) \ = < 0 > \cup \bigcup_{i=1}^{\nu(\omega)} I(\omega_i),$$

the proof of Eq. (B.3) is reduced to careful "bookkeeping" of the indicator functions $Z_\iota(y, t - l, \omega_i)$ and $Z_{i\iota}(y, t, \omega)$.

B.2.2 Generating functions

Let

$$F(s, t) = \sum_{r=0}^{\infty} \mathrm{P}\{Z(t) = r\} s^r. \tag{B.4}$$

Because the case $Z(t) = \infty$ has not yet been eliminated, it can be $F(1, t) < 1$. However, also in this case, the basic properties of the pgf's are verified. Let us note the alternative expression

$$F(s, t) = \mathrm{E}[s^{Z(t)}] \equiv \int_\Omega s^{Z(t,\omega)}\, dP(\omega), \tag{B.5}$$

where $0^0 = 1$ and $s^\infty = 0$, even if $s = 1$.

Theorem 32. *The generating function F satisfies the integral equation*

$$F(s, t) = s[1 - G(t)] + \int_{0-}^{t+} f[F(s, t - u)]\, dG(u), \tag{B.6}$$

where $t \geq 0$ and $s \in [0, 1]$.

Proof. Based on Eq. (B.5), let us write

$$F(s, t) = \int_\Omega s^{Z(t,\omega)}\, dP(\omega) = \int_{\{l>t\}} s^Z\, dP + \sum_{k=0}^{\infty} \int_{\{l \leq t, \nu = k\}} s^Z\, dP. \tag{B.7}$$

Because $Z(t, \omega) = 1$ if $l > t$, we have $\int_{\{l>t\}} s^Z \, dP = s \Pr\{l > t\} = s[1 - G(t)]$.

Let us consider Ω as a product space $\Omega' \times \Omega_1 \times \Omega_2 \times \cdots$ of points $(l, \nu; \omega_1, \omega_2, \ldots)$. Let P' be the probability measure on the pair (l, ν) and let P_i be the probability measure on Ω_i. Now, it is possible to use Eq. (B.3). If l is fixed, then $Z(t-l, \omega_i)$ is a function on Ω_i and hence, if k is any positive integer, we have

$$\int_{\{l\le t,\nu=k\}} s^Z \, dP = \int_{\{l\le t,\nu=k\}} dP'(l, \nu) \int_{\Omega_1} s^{Z(t-l,\omega_1)} \, dP_1 \ldots \int_{\Omega_k} s^{Z(t-l,\omega_k)} \, dP_k.$$

Now, each of the integrals $\int_{\Omega_i} s^{Z(t-l,\omega_i)} \, dP_i$ is equal to $F(s, t-l)$, as the probability measure $dP_i(\omega_i)$ is exactly the same as $dP(\omega)$. Hence, the last equation is equal to $p_k \int_{0-}^{t+} [F(s, t-u)]^k \, dG(u)$. The same can be seen directly true if $k = 0$. Substitution into the right-hand side of Eq. (B.7) yields the desired result.

B.2.3 Uniqueness of $F(s, t)$ and finiteness of $Z(t)$

Theorem 32 states that the pgf of $Z(t)$ satisfies Eq. (B.6), but it does not state that this solution is unique, nor that $\lim_{s\uparrow 1} F(s, t) = 1$ [i.e., that $Z(t) < \infty$]. We will outline here the arguments proving both these properties.

Regarding uniqueness, let us assume that there exists another pgf solution $\tilde{F}(s, t)$ of Eq. (B.6). Then,

$$|F(s, t) - \tilde{F}(s, t)| \le \int_0^t |F(s, t-y) - \tilde{F}(s, t-y)| \, dG(y). \qquad \text{(B.8)}$$

We see that because both F and $\tilde{F}$ are pgf's, $|F(s, t) - \tilde{F}(s, t)| \le 1$. Substituting into the right-hand side of Eq. (B.8), we obtain $|F - \tilde{F}| \le G(t)$. Substituting this and repeating the estimate, we obtain that $|F - \tilde{F}| \le G^{*i}(t)$ for any i. However, $\lim_{i\to\infty} G^{*i}(t) = 0$ for any t (see Lemma 4), which yields $|F - \tilde{F}| = 0$.

Finiteness of $Z(t, \omega)$ may be obtained by estimating another random variable $\bar{Z}(t, \omega)$ equal to the total number of objects in family $I(\omega)$ that are born up to and including time t (i.e., the counting function of births). Of course, $Z(t, \omega) \le \bar{Z}(t, \omega)$. We will consider the expected value of $\bar{Z}$. If it is finite, then $\bar{Z}$ is finite and so is Z [and, consequently, $F(1-, t) = 1$].

For the argument, let us consider an object $< \iota > \neq < 0 >$, where $\iota = i_1 i_2 \cdots i_k$. Let u_ι be a random variable that is 1 if $< \iota >$ is in the family $I(\omega)$ (i.e., if it is ever born), and 0 otherwise, and let v_ι be a random variable that is 1 if $l + l_{i_1} + \cdots + l_{i_1 i_2 \cdots i_{k-1}} \le t$ and 0 otherwise. Then, $< \iota >$ is born at or before t if and only if $u_\iota v_\iota = 1$, and

$$\bar{Z}(t) = 1 + \sum_{k=1}^{\infty} \sum_{i_1 i_2 \cdots i_k = 1}^{\infty} u_{i_1 i_2 \cdots i_k} v_{i_1 i_2 \cdots i_k}.$$

The expected value $\mathrm{E}(v_\iota)$ is equal to $G^{*k}(t)$. The rv u_ι is the indicator function of the event that object $< \iota >$ is ever born and, therefore, its expectation is equal to the probability of this event; that is, to

$$\mathrm{E}(u_\iota) = \mathrm{P}\{u \ge i_1, u_{i_1} \ge i_2, \ldots, u_{i_1 \cdots i_{k-1}} \ge i_k\}$$

$$= \mathrm{P}\{u \geq i_1\}\mathrm{P}\{u_{i_1} \geq i_2\}\ldots\mathrm{P}\{u_{i_1\cdots i_{k-1}} \geq i_k\}.$$

The u_ι's and v_ι's are independent, so that

$$\mathrm{E}[\bar{Z}(t)] = 1 + \sum_{k=1}^{\infty} G^{*k}(t) \sum_{i_1} \mathrm{P}\{u \geq i_1\} \sum_{i_2} \mathrm{P}\{u_{i_1} \geq i_2\} \ldots \sum_{i_k} \mathrm{P}\{u_{i_1 \ldots i_{k-1}} \geq i_k\}$$

$$= 1 + \sum_{k=1}^{\infty} G^{*k}(t)[f'(1-)]^k.$$

Lemma 4 states that this sum is $< \infty$ for all t and so $\mathrm{E}[\bar{Z}(t)] < \infty$.

APPENDIX C

General Processes

C.1 Introduction to the Jagers–Crump–Mode Process

This section is a useful reference, but it can be omitted at first reading. Its aim is to introduce the reader in an informal way to the basics of the general branching processes. In most part, the book is concerned with less general processes; therefore the subject can be postponed to a later reading. However, there are issues that are best expressed when phrased in terms of general processes. An example is an application of a general process to cell populations in Section C.2. Another recent example is an application of the general process in the genetics of aging (Olofsson et al. 2001). This latter work is also, to our knowledge, the only such application based on real-life data.

C.1.1 Definition of the general branching process

A basic source concerning general branching processes is the book by Jagers (1975). Our account is also based on Taïb (1992).

Individuals

We consider development in time of a population started by a single individual. The individuals can be considered elements of the set

$$I = \bigcup_{n=0}^{\infty} N^n,$$

called the Ulam–Harris space, where $N = \{1, 2, \ldots\}$ and $N^0 = \{0\}$. Individual 0 is the ancestor of the population. Each element of N^n is of the form $x = (x_1, \ldots, x_n)$. The meaning of this notation is that the individual belongs to the nth generation and is the x_nth progeny of the x_{n-1}st progeny, ..., of the x_1st progeny of the ancestor. This description is redundant, as not all these individuals will come to existence in a given realization of the process. Each of the individuals evolves in a space Ω, which is large enough to allow for all possible life spans and progeny-bearing processes of this individual. An element $\omega \in \Omega$, is this individual's life. The probability measure on a σ-algebra $\mathcal{F}$ of Ω is called Q.

Lives

For each individual, $\tau(\omega, k)$, $k = 1, 2, \ldots$, denotes successive ages at childbearing. In particular, $\tau(\omega, k)$ is the age at which the individual has its kth progeny. These ages are organized as epochs of a point process, a random collection of time moments or equivalently a random collection of non-negative integer-valued measures, denoted ξ. Mathematically,

$$\xi(\omega, [0, t]) = \xi(t) = \#\{k : \tau(\omega, k) \leq t\}$$

is the counting function of births (i.e., the number of progeny begotten before or at the age of t). In addition, λ, the duration of life ω of an individual, is a random variable $\lambda : \Omega \to R^+$.

The time evolution of the individuals is governed by the connections between their times of births. Let σ_x denote the moment of birth of individual x ($\sigma_0 = 0$, for the ancestor). Then, if we denote by xk the individual being the kth progeny of x, we set

$$\sigma_{xk} = \sigma_x + \tau_x(k).$$

In this latter expression, the argument ω is dropped, as it will be frequently done.

Construction of the process

If the space Ω is a Polish space (i.e., it is metric, complete, and separable), then the σ-algebra $\mathcal{F}$ can be selected as the class of Borel sets of Ω. The triplet (Ω, $\mathcal{F}$, Q) is the probability space of a single individual. If we assume that the lives of individuals are independent, then the space of the process can be constructed as a product space of the form $(\Omega^I, \mathcal{F}^I, Q^I)$, where I is the collection of all individuals. From now on, we will write P instead of Q^I and ω instead of $\{\omega_x, x \in I\}$.

The model presented can be specialized to include the classical branching processes, by assuming that all $\tau(\omega, k)$, $k = 1, 2, \ldots$, are concentrated at $\lambda(\omega)$ (i.e., all progeny are born at the same time). Then, if $\lambda(\omega) = 1$, we obtain the Galton–Watson process. If $\lambda(\omega)$ is a non-negative rv, we obtain the Bellman–Harris process and so forth.

C.1.2 Random characteristics and basic decomposition

The method of random characteristics makes it possible to account for individuals existing in the process, individuals being born during a given time interval, individuals with ages from a given interval, individuals with a given number of progeny and so forth. The random characteristic is a random function $\chi_x(a)$ defined on an individual's life. It defines the contribution, of a desired type, of individual x, from its birth until it reaches age a. The summary contribution of all individuals at a given time t, is equal to

$$Z_t^\chi = \sum_{x \in I} \chi_x(t - \sigma_x).$$

Z_t^χ is called the process counted by random characteristic $\chi_x(a)$. For example, if

$$\chi_x(a) = \begin{cases} 1 & \text{if } a \geq 0 \\ 0 & \text{otherwise,} \end{cases}$$

then Z_t^χ counts all individuals born until time t. If

$$\chi_x(a) = \begin{cases} 1 & \text{if } a \in [0, \lambda_x) \\ 0 & \text{otherwise,} \end{cases} \tag{C.1}$$

then Z_t^χ counts all individuals alive at time t. If

$$\chi_x(a) = \begin{cases} 1 & \text{if } a \in [0, \lambda_x) \cap [\tau_x(k), \infty) \\ 0 & \text{otherwise,} \end{cases}$$

then Z_t^χ counts all individuals alive at time t, with at least k progeny born before t.

For the process counted by random characteristics, it possible to write a backward decomposition, analogous to Eq. (1.1):

$$Z_t^\chi = \chi_0(t) + \sum_{i=1}^{X} Z_{t-\tau_0(i)}^{\chi\,(i)},$$

where X is the number of progeny effectively begotten by the ancestor and superscript (i) denotes the ith iid copy of the process.

C.1.3 Expectations, Malthusian parameter, and exponential growth

Reproductive measure is the expectation of the point process of progeny births,

$$\mu(A) = \mathrm{E}[\xi(\omega, A)].$$

It is characterized by the reproductive counting function $\mu(a) = \mu([0, a])$. The expectation of the process, $m_t = \mathrm{E}(Z_t^\chi)$ counted by characteristic $\chi(a)$ with expectation $g(a) = \mathrm{E}[\chi(a)]$, can be represented by the expression

$$m_t = \sum_{n=0}^{\infty} \int_0^t g(t-u)\, d\mu^{*n}(u) = \int_0^t g(t-u)\, d\nu(u),$$

where $\nu(u) = \sum_{n=0}^{\infty} \mu^{*n}(u)$. The nth convolution power of the reproductive measure, μ^{*n}, counts the expected number of progeny born to the nth-generation individuals in the process. Then, each of μ^{*n} has to be convolved with the expectation of the random characteristic, to account for proper bookkeeping, and the result summed over all generations of the process. Under mild conditions (e.g., no concentration of births at age 0 and expected total progeny of an individual finite), this sum is finite. Expectation m_t satisfies a renewal-type integral equation:

$$m_t = \int_0^t m_{t-u}\, d\mu(u) + g(t). \tag{C.2}$$

A major role in the theory is played by the Malthusian parameter, which determines (if it exists) the asymptotic rate of growth of m_t. The Malthusian parameter is the real solution of the equation

$$\hat{\mu}(\alpha) \equiv \int_0^\infty e^{-\alpha u}\, d\mu(u) = 1.$$

This solution, if it exists, is unique. In what follows, we will limit ourselves to the supercritical case {i.e., when $\mu([0, \infty)) > 1$} [see the classification (1.5)]. In this case the Malthusian parameter exists and is positive. The renewal theorem demonstrates, in the same way as was explained in Section 5.2 for the Bellman–Harris process, that m_t behaves asymptotically like $e^{\alpha t}$;

$$e^{-\alpha t} m_t \longrightarrow \frac{\int_0^\infty g(u) e^{-\alpha u}\, du}{\underbrace{\int_0^\infty u e^{-\alpha u}\, d\mu(u)}_{\beta}} \equiv c(\chi) \quad \text{as } t \to \infty. \tag{C.3}$$

If we assume that all progeny are born at the same time τ in the life of the individual, so that $\mu(u) = mG(u)$, where m is the mean count of progeny and $G(\cdot)$ is the cumulative distribution of τ, and that this is exactly the moment of individual's death (i.e., that $\lambda = \tau$), we obtain the Bellman–Harris process of Chapter 5. If we wish to account for individuals alive at time t, then we use the random characteristics of the form $\chi_x(a) = 1$ if $a \in [0, \tau)$, and $\chi_x(a) = 0$, otherwise, as in Eq. (C.1). This means that $g(u) = 1 - G(u)$. Substituting into expression (C.3), we obtain the expression derived for the Bellman–Harris process [Eq. (5.13)].

Without getting into more detail, we state that in the supercritical case, the entire process counted by a random characteristic behaves very much the same way as its expectation. Indeed, there exists a random variable W, with $\mathrm{E}(W) = 1$, such that

$$Z_t^\chi e^{-\alpha t} \longrightarrow c(\chi) W$$

as $t \to \infty$, with probability 1.

C.1.4 Abstract type spaces and composition of the process

Let us suppose that each newborn individual is endowed, at birth, with a type selected from a measurable space $(\Gamma, \mathcal{G})$, where $\mathcal{G}$ is a σ-algebra of subsets of Γ. In other words, there exist measurable mappings $\rho(j) : \Omega \to \Gamma$, which determine the types of newborn individuals. The point process ξ, which describes reproduction, is now defined by

$$\xi(A \times B) = \#\{i \in N; \rho(i) \in A, \tau(i) \in B\}.$$

Intuitively, $\xi(A \times B)$ is the number of progeny of an individual, born in time set B, with types in set A. The population of individuals can be defined on $(\Gamma \times \Omega^I)$, where Γ describes the type of the ancestor. The theorem of Ionesco–Tulcea enables one to construct a unique probability measure P_γ on $(\Gamma \times \Omega^I, \mathcal{G} \times \mathcal{A}^I)$ for the process with a type-γ ancestor. Similarly as before, a major role is played by the reproduction kernel $\mu(\gamma, A \times B) = \mathrm{E}_\gamma[\xi(A \times B)]$. For each real λ, we define

$$\mu_\lambda(\gamma, d\gamma' \times du) = e^{-\lambda u}\mu(\gamma, d\gamma' \times du)$$

and

$$\hat{\mu}_\lambda(\gamma, d\gamma') = \int_0^\infty \mu_\lambda(\gamma, d\gamma' \times du).$$

The Malthusian parameter α is selected so that the kernel $\hat{\mu}_\alpha(\gamma, d\gamma')$ has a Perron–Frobenius eigenvalue equal to 1 (assuming this latter exists). The Perron–Frobenius eigenvalue is the real eigenvalue strictly dominating absolute values of all remaining eigenvalues. If we set $\nu_\alpha(\gamma, d\gamma' \times du) = \sum_{n \geq 0} \mu_\alpha^n(\gamma, d\gamma' \times du)$, where $\mu_\alpha^n(\gamma, d\gamma' \times du)$ is the n-fold convolution of measure $\mu_\alpha(\gamma, d\gamma' \times du)$ with respect to elements $d\gamma' \times du$, we can write

$$\mathrm{E}_\gamma[e^{-\alpha t} Z_t^\chi] = \int_{\Gamma \times R_+} \mathrm{E}_\gamma[e^{-\alpha(t-u)}\chi(t-u)]\nu_\alpha(\gamma, d\gamma' \times du).$$

So, we see that $\mathrm{E}_\gamma[Z_t^\chi]$ is of the form $R * g(\gamma, t)$, where $R = \nu_\alpha$ and

$$g(\gamma, t) = \mathrm{E}_\gamma[e^{-\alpha t}\chi(t)].$$

Asymptotic behavior of the expectation of the process and of the process itself in the supercritical case ($\alpha > 0$) depends on the conservativeness of the kernel $\hat{\mu}_\alpha(\gamma, d\gamma')$. For countably generated $\mathcal{G}$, the kernel is conservative if its potential $\hat{\nu}_\alpha(\gamma, d\gamma') = \sum_{n \geq 0} \hat{\mu}_\alpha^n(\gamma, d\gamma')$ has the property that there exists a σ-finite measure m on $(\Gamma, \mathcal{G})$ such that

$$m(A) > 0 \implies \hat{\nu}_\alpha(\gamma, A) = \infty \tag{C.4}$$

for all $\gamma \in \Gamma$. This property is a generalization of positive regularity of matrices.

If the kernel $\hat{\mu}_\alpha$ is conservative, there exists an eigenfunction h satisfying

$$\begin{aligned} h(\gamma) &= \int_{R_+} \int_\Gamma e^{-\alpha u} h(\gamma')\mu(\gamma, d\gamma' \times du) \\ &= \int_\Gamma h(\gamma')\mu_\alpha(\gamma, d\gamma'). \end{aligned} \tag{C.5}$$

So, $e^{-\alpha u}[h(\gamma')/h(\gamma)]\mu(\gamma, d\gamma' \times du)$ has total mass on $\Gamma \times R_+$ equal to 1 and it is a probability measure. $h(\gamma)$ is the reproductive value of individuals of type

γ. It indicates the relative long-term contribution of individuals of this type to the population.

If the kernel $\hat{\mu}_\alpha$ is conservative, there also exists a probability measure π, which satisfies

$$\pi(d\gamma') = \int_\Gamma \hat{\mu}_\alpha(\gamma, d\gamma')\pi(d\gamma). \tag{C.6}$$

This equation can also be written in the following manner:

$$h(\gamma')\pi(d\gamma') = \int_\Gamma \frac{h(\gamma')}{h(\gamma)}\hat{\mu}_\alpha(\gamma, d\gamma')h(\gamma)\pi(d\gamma)$$

if $\inf h(\gamma) > 0$. We can then normalize the equation so that we obtain $\int_\Gamma h(\gamma)\pi(d\gamma) = 1$. The measure π, defined above, can be interpreted as a stable distribution of the types of the newborn. Consequently, an individual drawn at random from a very old population is of a random type decided by π, independently of the initial conditions.

Another interesting expression,

$$\beta = \int_\Gamma \int_\Gamma \int_{R_+} te^{-\alpha t}h(\gamma')\mu(\gamma, d\gamma' \times dt)\pi(d\gamma),$$

can be considered the expected age at reproduction.

Similarly as in the single-type case, in the supercritical case ($\alpha > 0$) a generalization of the key renewal theorem makes it possible to calculate the limit of $\mathrm{E}[e^{-\alpha t} Z_t^\chi]$. We will denote $\mathrm{E}_\pi(X) = \int_\Gamma \mathrm{E}[X]\pi(d\gamma)$, the expectation in the process with the type of ancestor being randomly drawn according to measure π. Then, we have

$$\mathrm{E}[e^{-\alpha t} Z_t^\chi] \longrightarrow \frac{\mathrm{E}[\hat{\chi}(\alpha)]}{\alpha\beta}h(\gamma)$$

as $t \to \infty$, for all γ except sets of π-measure 0. The process behaves in the supercritical case very much like its expectation (also, see Berndtsson and Jagers 1979).

The multitype formulation provides a great generality and was used in applications, particularly concerning evolution theory (Taïb 1992).

C.2 *Application*: Alexandersson's Cell Population Model Using a General Branching Process

An elegant example of modeling using general processes and counting characteristics (Section C.1) is a part of Alexandersson's (1999) thesis. This application demonstrates how a branching process approach complements existing deterministic approaches while the construction of the process is very straightforward.

C.2.1 The model

Let us consider a cell population, where each cell inherits a type at birth, grows during a stochastic time span, and when its cell cycle is completed, it divides into two not necessarily equal daughter cells. The type of the individual is the birth size of the cell expressed as mass, volume, DNA content, and so forth. Because cells have only two progeny, the Ulam–Harris space of all possible cells reduces to

$$I = \bigcup_{n=0}^{\infty} \{1, 2\}^n,$$

where $\{1, 2\}^0 = \{0\}$.

The type space is an interval $S = (0, M]$ of the real line, where $M < \infty$ is the largest possible birth size of a cell and $\mathcal{S}$ is the Borel-σ-algebra on S. A cell with birth size $r \in S$ chooses a life ω from $(\Omega, \mathcal{A})$ using $P(r, \cdot)$, the life law of cells of type r.

We construct the population space $(S \times \Omega^{\mathcal{I}}, \mathcal{S} \times \mathcal{A}^{\mathcal{I}})$ as in Section C.1. Under the assumption that the daughter processes of different cells are conditionally independent, there exists a unique probability measure P_r on the entire population process, where $r \in S$ is the type of the ancestor.

The size of a cell with initial size r increases with time according to a deterministic growth function g. We let $m(r, t)$ denote the size of an r-type cell at age t. The functions m and g are related by the initial value problem

$$\frac{dm}{dt} = g(m), \qquad m(r, 0) = r.$$

The cell grows and, after division, the daughter cells do not necessarily have the same size (type) at birth. Note that we do not allow cell death in this model, so our branching population is supercritical. Let λ denote the age of the cell at division (the cell cycle time) and let the distribution of λ be defined by its hazard rate function $b(s)$, $s \in (0, 2M]$ {i.e., $\mathrm{P}[\lambda > s] = \exp[-\int_0^s b(u)\, du]$}.

A cell of type r divides into fractions δ and $1 - \delta$, where δ is a random variable on $(0, 1)$ with density function $f_\delta(m, p)$, $p_1 \le p \le p_2$, where $p_1 = 1 - p_2 \in (0, 1)$ depends on $m = m(r, \lambda)$, the cell size at division. We will assume that f_δ is unimodal and that δ is symmetrically distributed around $1/2$ {i.e., for all $r \in S$, $f_\delta(m, p) = f_\delta(m, 1 - p)$, and $\mathrm{E}_r[\delta] = 1/2$}.

Let $T(x) = \int_0^x [1/g(y)]\, dy$, $x \in S$. To see how to interpret this function, consider

$$T(x) - T(r) = \int_r^x \frac{1}{g(y)}\, dy. \tag{C.7}$$

Making a change of variable $y = m(r, t)$ yields $dy = dm(r, t) = g(m(r, t))\, dt$ and Eq. (C.7) becomes

$$\int_0^u \frac{g(m(r, t))}{g(m(r, t))}\, dt = \int_0^u dt = u,$$

where u is the time it takes for a cell to grow from size r to size x. Consequently, $T(x) - T(r)$ is precisely this time. Because $T(m(r, t)) - T(r) = t$, we have $m(r, t) =$

$T^{-1}(T(r)+t)$. Further, let $C(x) = \int_0^x [b(y)/g(y)]\, dy$ and $Q(x) = b(x)/[xg(x)]$, and assume that each cell has to divide before it reaches size $2M$; that is, b is such that for $\epsilon > 0$,

$$\int_0^{2M} \frac{b(y)}{g(y)}\, dy = \infty \quad \text{and} \quad \int_0^{2M-\epsilon} \frac{b(y)}{g(y)}\, dy < \infty.$$

The reproduction kernel $\mu(r,\ ds \times dt)$, which is the expected number of children with birth sizes in ds to a cell of type r with age in dt, takes the form

$$\begin{aligned} \mu(r, ds \times dt) &= \mathrm{E}_r[\xi(ds \times dt)] \\ &= \mathrm{E}_r[\mathbf{1}(\lambda \in dt)(\mathbf{1}(\delta m(r,\lambda) \in ds) + \mathbf{1}((1-\delta) m(r,\lambda) \in ds))] \\ &= 2\int_0^\infty \mathbf{1}(u \in dt) \int_0^1 \mathbf{1}(pm(r,u) \in ds) f_\delta(m(r,u), p)\, dp \\ &\qquad \times b(m(r,u)) \exp\left[-\int_0^u b(m(r,v))\, dv\right] du, \end{aligned}$$

where the factor 2 comes from the fact that δ and $(1-\delta)$ are identically distributed.

The inner integral is zero everywhere except when $p = s/m(r,u)$ and $dp = ds/m(r,u)$, so we have

$$\begin{aligned} \mu(r, ds \times dt) = 2\int_0^\infty & \mathbf{1}(u \in dt) f_\delta(m(r,u), s/m(r,u)) \frac{b(m(r,u))}{m(r,u)} \\ & \times \exp\left[-\int_0^u b(m(r,v))\, dv\right] du\, ds. \end{aligned} \tag{C.8}$$

Making a change of variable in the same manner as above, with $x = m(r,u)$, we get that $du = [dx/g(x)]$ and the kernel becomes

$$\mu(r, ds \times dt) = 2\int_r^{2M} \mathbf{1}(T(x) - T(r) \in dt) f_\delta(x, s/x) Q(x) e^{-(C(x)-C(r))}\, dx\, ds.$$

C.2.2 Existence of the stable birth size distribution

If the Malthusian parameter α exists such that $\hat{\mu}_\alpha$ is conservative, then the Perron–Frobenius theorem gives the existence of a function h [see Eq. (C.5)] and a measure π [see Eq. (C.6)]. By requiring a strong or positive α-recurrence (Jagers and Nerman, 1996) and $\inf h > 0$ we can norm to

$$\int_S h(s)\pi(ds) = 1, \qquad \int_S \pi(ds) = 1.$$

The measure π is then called the stable-birth-type-distribution. Hence, we want to prove the existence of the Malthusian parameter [i.e., prove the existence of a number $\alpha > 0$ such that the Perron root $\rho(\hat{\mu}_\alpha) = 1$], where

$$\hat{\mu}_\alpha(r, A) = \int_{\mathrm{R}_+} e^{-\alpha t} \mu(r, A \times dt)$$

and also that $\hat{\mu}$ is conservative.

Theorem 33. *Under the assumptions stated in Section C.2.1 on the reproduction kernel μ, the Malthusian parameter α exists and the kernel $\hat{\mu}_\alpha$ is conservative.*

C.2.3 Asymptotics of the cell model

We discuss the asymptotics of our cell model. When looking at a population, one can either consider all cells alive at the moment or all cells born into the population up until now. Even if it seems more natural to look at all cells alive, it is mathematically more convenient to consider all born. In this chapter, we will concentrate on all born cells, but we will also show that all the results presented can easily be obtained for all cells alive as well. When calculating the asymptotics of our model, we construct random characteristics used to count the population with respect to some property. An alternative way, described in Jagers and Nerman (1996) is to sample an individual at random in an already stabilized population and consider the population with time centered around this individual. The individual sampled at random is called ego.

The α curve is the graph of the function $\alpha(a)$ describing the proportion of cells still undivided at age a. An alternative interpretation is that $\alpha(a)$ is the probability that the age at division of a cell sampled at random, ego is larger than a. In order to find an expression for $\alpha(a)$, we define a random characteristic χ (cf. Section C.1) such that z_t^χ counts the number of cells born up to time t with respect to χ. Then, if y_t denotes the number of all cells born up to time t, we can use the result that under suitable conditions

$$\frac{z_t^\chi}{y_t} \to \mathrm{E}_\pi[\hat{\chi}(\alpha)] \quad \text{as } t \to \infty$$

in probability (on the set of nonextinction), where $\mathrm{E}_\pi[X] = \int_S \mathrm{E}_s[X]\pi(ds)$, $\hat{\chi}(\alpha) = \int_{\mathrm{R}_+} \alpha e^{-\alpha t}\chi(t)\,dt$, and π is the stable-birth-type distribution.

The random characteristic that gives score one for each cell x born up to time t and with life length λ_x longer than a can be written as

$$\chi_x(t) = \mathbf{1}_{\mathrm{R}_+}(t - \tau_x)\mathbf{1}(\lambda_x > a),$$

where τ_x is the birth time for cell x. Making a change of variable $u = t - \tau_x$ gives

$$\chi(u) = \mathbf{1}_{\mathrm{R}_+}(u)\mathbf{1}(\lambda > a).$$

This yields

$$\begin{aligned}
\alpha(a) = \mathrm{E}_\pi[\hat{\chi}(\alpha)] &= \int_S \mathrm{E}_r[\hat{\chi}(\alpha)]\pi(dr) \\
&= \int_S \mathrm{E}_r\left[\int_{\mathrm{R}_+} \alpha e^{-\alpha u}\chi(u)du\right]\pi(dr) \\
&= \int_S \int_{\mathrm{R}_+} \alpha e^{-\alpha u}\,du\,\mathrm{E}_r[\mathbf{1}(\lambda > a)]\pi(dr) \\
&= \int_S \mathrm{P}_r(\lambda > a)\pi(dr)
\end{aligned}$$

$$= \int_S \exp\left[-\int_0^a b(m(r, v))\, dv\right] \pi(ds).$$

The β curves are used to describe the proportions of sister cells, cousin cells, and so on with life lengths that differ by more than a time units. The β_1 curve describes this proportion for sister cells, β_2 for cousin cells, and so on. Alexandersson's (1999) thesis includes further asymptotic results for the β curves and numerical computations for the model we outlined. Furthermore, it also deals with a much more complicated example of cell proliferation, which we consider, using different methods, in Section 7.7.2.

APPENDIX D

Glossaries

D.1 Biological Glossary for Mathematicians

Cross-references to other glossary terms are *italicized.*

Amino acids The 20 different basic units of *proteins.*

Amplification (Gene Amplification) The increase in the number of copies of a *gene*. May result from errors in DNA *replication* or *recombination.*

Antibody A *protein* produced by the immune system in response to a foreign molecule (*antigen*) that interacts specifically with the foreign molecule.

Antigen A molecule that induces an *antibody.*

bp Base pair(s), usually used as a unit of length of a *DNA* strand, spanning one pair of complementary nucleotides.

Bacteria *Cells* of a lower form of life without a *nuclear* membrane.

Cancer A population of *cells* that continue to divide and survive under conditions in which normal cells would stop dividing or die. The cancer cell population usually is initiated from a single cell (clonal origin). As the progeny of the single cell multiply they accumulate *mutations* and acquire new characteristics (tumor progression). They may invade adjacent tissues and travel to distant sites to form secondary tumors (metastases).

Cell The basic unit of life. Cells of higher forms of life have an outer membrane surrounding the cytoplasm and the *nucleus*. In the cytoplasm there are *proteins* (enzymes) that carry out biochemical functions, machinery (ribosomes) for making proteins, and compartments (organelles) such as *mitochondria*. Higher forms of life, such as mammalian cells, which have a membrane surrounding their nucleus, are referred to as eukaryotes. Lower forms of life, such as *bacteria*, which do not have a membrane surrounding their nucleus, are referred to as prokaryotes.

Cell cycle The stages of *cell* growth and division. Includes the following stages (phases): division of one cell to produce two cells (cytokinesis), a gap of time (G_1

phase) between cytokinesis and the initiation of *DNA* synthesis (S phase), a gap of time (G_2 phase) between the end of DNA synthesis and the formation of visible chromosomes, and *mitosis* (M phase). In mitosis, the duplicated *chromosomes* (chromatids) containing replicated *DNA* are partitioned to new *cells* at cell division. The time between one cell division and another is referred to as the cell lifetime.

Centromere A part of the *chromosome* required for proper movement of the daughter *chromosomes* (chromatids) to daughter *cells*. A piece of *DNA* that is not part of a chromosome and does not contain a centromere DNA sequence is referred to as an acentric extrachromosomal element or double minute chromosome. Such acentric extrachromosomal elements do not segregate properly into daughter cells.

Chemotherapy The treatment of *cancer* cells with chemicals that kill them. In combination drug therapy, two or more chemicals with different modes of action are used to increase the efficiency of killing cancer cells.

Chromosome The linear structure containing *DNA* and *protein* that is visible under a microscope at *mitosis*. Chromosomes contain DNA sequences (*genes*) that code for proteins and DNA sequences that do not code for proteins. Among the noncoding DNA sequences, there are *centromeres* necessary for the separation of daughter chromosomes (chromatids) during *mitosis* and *telomeres*, which function to maintain the integrity of the ends of chromosomes.

Colony A population of *cells* that are the progeny of a single cell.

DNA Deoxyribonuleic acid; the genetic material. A long double helix with a structure similar to a twisted ladder. The backbones of the ladder are strands composed of alternating sugar (deoxyribose) and phosphate groups. The rungs of the ladder are pairs of nucleotide subunits. The nucleotide subunits are abbreviated A, T, G, and C. A is paired with T, and G is paired with C. The genetic information in DNA is stored in the sequence of nucleotides. The information is transcribed into complementary copies of a sequence of nucleotides in messenger *RNA* and is then translated into a sequence of *amino acids* in *protein*. During DNA replication, the two strands of a double helix separate and each acts as a template to synthesize a new complementary strand. Each of the two double helices (one new strand and one old strand) is contained in each one of a pair of sister chromatids (the daughters of *chromosomes*). The sister chromatids segregate into daughter cells at *mitosis*.

Drug resistance The continued survival of *cells* in the presence of chemicals (drugs) intended to kill them. Resistance to two or more drugs is referred to as double resistance or cross-resistance.

Eve The hypothetical common human female ancestor of all extant humans. Suggested by some common genetic features of individuals in current human populations.

Flow cytometry A method for the analysis of the distribution of the amount of a molecule (such as *DNA* or *protein*) in a population of *cells*. Cells are stained and pumped through a thin tube between a light source and a detector. Measurements of the amount of DNA per cell are used to indicate the number of cells in each phase of the cell cycle. Measurements of the amount of a specific protein per cell are used to indicate overproduction of the protein as a result of, for instance, *gene amplification*.

Fluctuation analysis Also, Luria and Delbrück fluctuation analysis. A method to determine *mutation* rates of *bacteria* or mammalian cells. Parallel cultures of cells are grown for a number of generations and then the number of mutants in each culture, the average number of mutants per culture, and the number of cultures containing no mutants are determined. This information can be used to calculate the number of mutations per cell per generation.

Gene A sequence of bases in *DNA* that codes for a *protein* and influences the inherited characteristics of a *cell* or organism.

Genome All of the *DNA* in an organism, including the DNA that codes for *proteins* and the DNA that does not code for proteins.

Heterogeneity (Tumor Heterogeneity) Populations of *cancer* cells that contain subpopulations with different characteristics, such as relative resistance to drugs.

Meiosis The formation of gametes (sex *cells*) by two successive cell divisions and only one round of *DNA* synthesis. This results in the segregation of nonidentical forms of genes (alleles) into different gametes. The gametes are haploid, containing half as much DNA as diploid body cells.

Mitochondria *Organelles* in the cytoplasm of *cells* of higher organisms needed for generating energy. Mitochondria contain *DNA*. They are inherited only from the mother, hence the term "maternal inheritance."

Mitosis The stage of the cell cycle of somatic (body) *cells* in which replicated *chromosomes* (chromatids) are separated into daughter cells. The result of mitosis is two daughter cells that have identical sets of *genes*. Daughter cells may be different in size as a result of asymmetric division of the cytoplasm at cell division.

Molecular clock hypothesis The assumption that *mutations* in a *gene* occur randomly and at an approximately equal rate over long time intervals during evolution.

Mutant An organism or *cell* that has a different inherited characteristic than the remainder of the cells in a population. Usually the result of a change in *DNA* sequence.

Mutation A change in *DNA* sequence. Usually detected by a sudden and inherited change in an observed characteristic (*phenotype*) of a *cell* or of an organism. However, a mutation may be detected directly by determining a change in the DNA sequence, even though there is no visible characteristic change in the cell or organism. The progeny of the mutant may revert to the previous phenotype, in which case the new mutation is referred to as a reverse mutation or back mutation. A phenotype resulting from a series of two mutations is referred to as a two-stage mutation. The rate of mutation may be determined by *fluctuation analysis*.

Nucleus The part of a *cell* containing *DNA*. The part of the cell outside of the nucleus is referred to as the cytoplasm.

Oncogene A *gene* (*DNA* sequence) associated with *cancer*. An oncogene can be detected and mapped by its pattern of inheritance in cancer-prone families. A piece of DNA containing an oncogene can be detected by the ability of the DNA to induce cancer-like changes when transferred into *cells* growing in culture.

Organelle A part of a *cell* which carries out a specialized function. An example is a mitochondrion (plural: *mitochondria*). A mitochondrion is a *DNA*-containing, membrane-enclosed structure located in the cytoplasm. It functions to produce high-energy molecules for cell metabolism. During cell division, mitochondria may or may not be distributed to daughter cells in equal numbers.

Phenotype The visible characteristics of a *cell* or organism; as opposed to genotype, the genetic information of a cell.

Plasmid In *bacteria*, a circular piece of *DNA* that is separate from the major (chromosomal) piece of DNA. Plasmids replicate and segregate at *cell* division independently of the chromosomal DNA. Each bacterial cell may contain multiple numbers of plasmids which may be randomly distributed at cell division.

Polymerase chain reaction An experimental procedure for obtaining a large number of copies of a piece of *DNA*. The procedure employs short pieces of DNA complementary to the ends of the desired sequence and the enzyme DNA polymerase to exponentially increase the number of copies of the desired DNA sequence.

Protein A polymer molecule consisting of monomer subunits of *amino acids*. The linear sequence of amino acids in a protein is determined by the corresponding sequence of nucleotides in *DNA* (*gene*). Some proteins (enzymes) function to encourage chemical reactions; other proteins have a structural function.

Quiescence A phase when *cells* are pausing before the initiation of *DNA* synthesis rather than actively progressing through the *cell cycle*. Most cells of higher organisms are quiescent rather than actively dividing.

Recombination The formation of new combinations of *genes* by the exchange of genetic information between *chromosomes*.

Repeat DNA Sequences of *DNA* nucleotides that are tandemly iterated. In some diseases, the number of repeats may vary between individuals, and the number may change from parents to progeny.

Replication The duplication of *DNA*. Two strands of DNA separate, like a zipper, at a moving replication fork. Each strand acts as a template to code for a complementary sequence of nucleotides in a new strand. The result is two new pieces of DNA, each double stranded, each piece containing one new strand and one old strand. This is referred to as semiconservative replication. Errors may occur during DNA replication, slippage at the replication fork or redundant replication forks, resulting in sequences that are added or deleted (*amplification* or deamplification).

RNA Ribonucleic acid. A molecule similar to *DNA*, but with a different sugar (ribose rather than deoxyribose), one different nucleotide (U instead of T), and mostly single stranded (rather than double stranded). There are several kinds of RNA. One of these, messenger RNA (mRNA), is transcribed as a complementary copy of the sequence of nucleotides in DNA and functions to determine the sequence of *amino acids* in *protein*.

Segregation The separation of different forms of *genes* (alleles) into sex *cells* (gametes) at *meiosis*. Also, the distribution of double minute *chromosomes* to daughter cells during *mitosis*.

Senescence The inability of some normal *cell* populations to continue to divide indefinitely when grown in culture. Some *cancer* cell populations can continue to divide indefinitely in culture and are therefore referred to as immortal. Senescence has been related to the continued activity of molecules that control *cell cycle* progression and to the maintenance of the length of *telomeres* at the ends of *chromosomes*.

Telomeres The ends of *chromosomes*. The *DNA* at the ends of chromosomes contains repeated sequences (terminal restriction fragments, TRF) that are necessary for replicating DNA at the ends of chromosomes and for maintaining the structural integrity of chromosomes.

Virus An intracellular parasite of *cells*. There are viruses of bacteria and of higher cells, including mammalian cells. They replicate within cells and can be transferred between cells. The extracellular forms contain genetic material (*DNA* or *RNA*), *proteins*, and some contain membranes. Within cells, the viral genetic material may subvert the machinery of the host cells and alter the host cell's properties. The genetic material of some viruses will actively replicate within a cell and produce new viruses. The genetic material of other viruses will integrate a DNA copy into the DNA of the host cell and replicate the viral genetic material along with the DNA of the host cell once per *cell cycle*.

D.2 Mathematical Glossary for Biologists

Cross-references to other glossary terms are *italicized*.

Abel's equation One of the classical functional equations of calculus. For a supercritical *branching process*, the characteristic function of the limit *random variable* W equal to the standardized particle count satisfies Abel's equation (3.19).

Age-dependent branching process A *branching process* in which the lifetimes of particles are non-negative *random variables*. In the special case when the lifetimes are exponentially distributed, the number of particles existing in the process, as a function of time, is a time-continuous *Markov chain*.

Asymptotic behavior Behavior of a time-dependent process (or a biological or physical phenomenon) after a sufficiently long time.

Backward approach Decomposition of the *branching process* into subprocesses started by direct progeny of the ancestor. By the branching property (a form of *self-recurrence*), these latter are distributed identically as the whole process. This decomposition provides the means to derive recurrent relationships or equations for the distributions of the process.

Bellman–Harris branching process A *branching process* in which the lifetimes of particles are non-negative *random variables* (age-dependent process) and the progeny is born exactly at the moment of the death of the parent.

Branching diffusion process A *branching process*, with a continuum *type space*, in which the type of the particle is defined as its position in a subset of real numbers (or points in higher-dimensional space) and the transitions in the type space are translations by a real-valued *random variable* (or a vector), with special rules on the boundary. The type may be understood as a spatial coordinate of the particle.

Branching process A random collection of individuals (particles, objects, cells), proliferating according to rules involving various degrees of randomness of the life length and the number of progeny of an individual. The unifying principle is the so-called branching property, which states that the life length and type of progeny of a newborn particle, conditional on the current state of the process, are independent of any characteristics of other particles present at this time or in the future. The branching property is a form of *self-recurrence*.

Branching random walk A *branching process*, with a *denumerable*-type space, in which the type of the particle is defined as its position in the set of integers (or non-negative integers) and the transitions in the type space are translations by an integer *random variable*, with special rules on the boundary. An example is the process of *gene amplification* in proliferating *cells*. In this process, the type of cell is the number of copies of a *gene* present in the cell's *DNA*. Progeny cells may gain or lose copies of this gene, inherited from the parent cell. So, if the number of gene copies in the parent is equal to i; then in the progeny, it may be equal to $i - 1, i,$ or $i + 1$.

Càdlàg path Function of time continuous from the right and bounded from the left of each point (French: **c**ontinue **à d**roite, **l**imitée **à g**auche).

Chapman–Kolmogorov equation Fundamental relationship governing the time evolution of *Markov chains*. It is represented in various forms {e.g., $P(s+t) = P(s)P(t)$ or $P_{ij}(s+t) = \sum_k P_{ik}(s)P_{kj}(t)$, where $P(s) = (P_{ij}(s))$ is the matrix (finite or infinite) of transition probabilities between states, $P_{ij}(s) = \mathrm{P}[X_{t+s} = j|X_t = i]$}. Intuitively, to calculate the probability of the chain moving from i to j in time $t + s$, it is necessary to add the probabilities of moving from i to k in time t and from k to j in time s, over all states k.

Criticality *Branching process* is critical if the expected (mean) count of progeny of a particle is equal to 1. It is supercritical if the mean count of progeny of a particle is greater than 1 and subcritical if it is less than 1. This classification leads to profound differences in *asymptotic properties* of the process. In particular, critical processes behave in a counterintuitive way because they become *extinct* with probability 1 while the expected number of particles stays constant.

Denumerable A set is called denumerable (or countable) if it is infinite but its elements can be indexed by non-negative integers. Other categories of infinite sets include continuum (i.e., a set the elements of which can be indexed by real numbers from an interval). The set of all rational numbers (ratios of integers) is countable; the set of all infinite sequences of zeros and ones is a continuum (because such sequences are just binary expansions of real numbers from the [0, 1] interval).

Exponential Steady State For idealized populations growing without spatial or nutritional constraints, the condition in which the number of individuals increases or decreases exponentially while the proportions of individuals in distinct age classes and any other identifiable categories remain constant. Usually attained *asymptotically*.

Extinction The event of all particles (individuals) of the *branching process* dying out.

Forward approach An approach dual to the backward approach, easiest to explain for the *Galton–Watson branching process*. Particles existing in generation t of the process are traced to their parents in generation $t-1$. Therefore, if the number Z_{t-1} of particles in generation $t-1$ is known, the number Z_t of particles in generation t is equal to the sum $Z_t = X_1 + X_2 + \cdots + X_{Z_{t-1}}$, where X_k is the number of progeny of the kth out of Z_{t-1} particles of generation $t-1$. This leads to a recurrence for the *pgf*'s of the particle counts.

Galton–Watson branching process Arguably, the simplest *branching process*. It evolves in discrete time measured by non-negative integers. At time 0, an ancestor individual (particle, cell, object) is born. At time 1, the ancestor dies, producing a random number of progeny. Each of these becomes an ancestor of an independent subprocess, distributed identically as the whole process. This definition implies that the numbers of progeny produced by each particle ever existing in the process are independent identically distributed *random variables* and that all particles live for one time unit. Discrete-time moments coincide with generations of particles. The number of particles existing in the Galton–Watson branching process, as a function of time, constitutes a time-discrete *Markov chain*.

Gelation In a model of aggregation of chemical molecules, the idealized process of infinite aggregation, resulting in disappearance of finite aggregates of molecules. In Macken and Perelson's branching model of aggregation, gelation is represented by escape of the *branching process* to infinity (possible only in the supercritical case).

Genealogies Branching (tree-like) graphs, usually random with respect to structure and branch lengths, representing ancestry of a sample of individuals from a *branching process* or, more generally, from an abstract or real-life population of molecules, *genes*, *cells*, or other objects. The process of reducing the number of distinct ancestors of the sample, followed in the reverse time, is called coalescence.

Genetic distance Distance between biological organisms, computed based on genetic characteristics. An example is the distance between relevant subsequences of *DNA* of the two individuals, computed as the number of nucleotides different in these two individuals (number of mismatches). For example, if in individual 1 the DNA sequence is ATGGACGA and in individual 2 it is ATcGgCGt, then the genetic distance is equal to 3.

iid Independent, identically distributed (*random variables*). The most frequently encountered assumption concerning a family of random variables. Makes proofs of theorems easier, when it can be assumed. In statistics, the so-called random samples are assumed to be iid.

Instability of branching processes The fact that, as time tends to infinity, the *branching process* either becomes *extinct* or infinitely large. Instability is due to the independence assumptions inherent in the definition of a branching process (i.e., that the number of progeny and life length of a newborn particle, conditional on the current state of the process are independent of any characteristics of other particles present at this time or in the future).

Jagers–Crump–Mode process The general *branching process*. The difference with respect to the classical branching processes, such as the *Galton–Watson branching process* or the *Bellman–Harris branching process* is that in the general process, the progeny may be produced before the death of the individual. The ages at which the individual begets progeny are random. Also, the *type space* may be of a very general form. The theory, developed for general processes, allows finding distributions of the process counted by random characteristics [i.e. of the weighted counts of events associated with a desired subclass of individuals (e.g., the number of first-born progeny of all individuals born after January 1, 1980, etc.)].

Kolmogorov theorem In the theory of *stochastic processes*, a fundamental result ensuring the existence of the stochastic process, given that for all finite collections of times, there exist joint distributions of *random variables*, being the values of the process at these times. These finite-dimensional distributions have to satisfy consistency conditions.

Linear-fractional case An important case of the *Galton–Watson branching process*, in which the number of progeny of an individual is a random variable with modified geometric distribution {i.e., $\mathrm{P}[X = 0] = 1 - bp/(1 - p)$ and $\mathrm{P}[X = k] = bp^k$, for $k = 1, 2, \ldots$. The name is derived from the fact that the *pgf* of such random variable is a ratio of two linear functions. In the linear-fractional case, the number of particles existing at any time has a modified geometric distribution, with parameters, which can be explicitly computed.

Malthusian parameter For a *branching process*, a parameter α such that the number $Z(t)$ of particles present in the process, normalized by dividing it by $\exp(\alpha t)$, converges to a limit *random variable*, as time tends to infinity. The Malthusian parameter always exists for the supercritical processes and is positive in this case.

Markov branching process A type of time-continuous *branching process*. At time 0, an ancestor individual (particle, cell, object) is born. The ancestor lives for time τ, which is an exponentially distributed *random variable*, and then the ancestor dies, producing a random number of progeny. Each of these becomes an ancestor of an independent subprocess, distributed identically as the whole process. The number of particles existing in the Markov branching process, as a function of time, is a time-discrete Markov chain (hence the name). Interestingly, if the Markov branching process is observed at times equal to multiples of a constant interval Δt, the numbers of particles at these observation times are distributed identically as in a *Galton–Watson branching process*.

Markov process A *stochastic process* with a limited memory (the Markov property). Intuitively, given the state of the process at time t, the future of the process depends only on this state and not on its states at times before t (time can be discrete or continuous). Mathematically,

$$\mathrm{P}[X_{t+s} \in A|X_s = x_s,\ 0 \leq s \leq t] = \mathrm{P}[X_{t+s} \in A|X_t = x_t],$$

where A is a subset of the state space of the process (space of values assumed by the process). The probability $P(s; x \to A) = \mathrm{P}[X_{t+s} \in A|X_t = x]$ is the transition probability from state x to set of states A, in time s. If the states of the process form a finite or *denumerable* set, then the process is called a Markov chain. In this case, it is possible to define a matrix (finite or infinite) of transition probabilities between states $P(s) = (P_{ij}(s))$, where $P_{ij}(s) = \mathrm{P}[X_{t+s} = j|X_t = i]$. For discrete time, $P(s) = P(1)^s$, where $P(1)$ is the single-step transition probability matrix. For continuous time (under some additional assumptions if the number of states is infinite), $P(s) = \exp(Qs)$, where Q is called the transition intensity matrix.

Martingale In the discrete-time case, a *stochastic process*, having the property that its expected value at time $t+1$, conditional on its values at all times before $t+1$, is equal to the process value at time t. Mathematically, $\mathrm{E}(X_{t+1}|X_1, X_2, \ldots, X_t) = X_t$. Martingales, under some additional conditions, converge to limits (which are *random variables*). For this reason, proving that a process is a martingale allows an insight into its *asymptotic behavior*. Continuous-time martingales behave in a similar way, but they are technically more involved.

Maximum likelihood Statistical methodology of estimating parameters of models, based on observations. It consists of expressing the probability of observations as a function of parameters. This function is known as the likelihood function, $L(\theta) = f_X(x; \theta)$, where $f_X(\cdot)$ is the density of the distribution of *random variable* X, x is the vector of observations of random variable (known), and θ is the vector of parameters of the distribution (unknown). The values of parameters, which maximize $L(\theta)$, are called the maximum likelihood estimates of the parameters and are denoted $\hat{\theta}$.

Moments Expected values of powers of a *random variable* X. Absolute moments of order k (or kth absolute moments) are defined as $\mathrm{E}(X^k)$, central moments as $\mathrm{E}\{[X-\mathrm{E}(X)]^k\}$, and factorial moments as $\mathrm{E}[X(X-1)(X-2)\cdots(X-k)]$. The first absolute moment, $\mathrm{E}(X)$, represents the central tendency of the random variable, the second central moment, $\mathrm{Var}(X) = \mathrm{E}\{[X-\mathrm{E}(X)]^2\}$, represents the dispersion of the random variable around the expected value.

Multitype Galton–Watson process (positive regular) Generalization of the usual (single-type) *Galton–Watson branching process*. It evolves in discrete time measured by non-negative integers. Each individual belongs to one of a finite number of types. At time 0, an ancestor individual (particle, cell, object), of some type, is born. Processes started by individuals of different types are generally different. At time 1, the ancestor dies, producing a random number of progeny of various types. The distribution of progeny counts depends on the type of parent. Each of the first-generation progeny becomes an ancestor of an independent subprocess, distributed identically as the whole process (modulo ancestor's type). In the multitype process, asymptotic behavior depends on the matrix of expected progeny count. Rows of this matrix correspond to the parent types and columns correspond to the progeny types. The largest positive eigenvalue of this matrix (the *Perron–Frobenius* eigenvalue) is the *Malthusian parameter* of the process, provided the process is supercritical (the Perron–Frobenius eigenvalue larger than 1) and positive regular. This latter means that parent of any given type will have among its (not necessarily direct) descendents, individuals of all possible types, with nonzero probability.

Parsimony method in phylogenetics A method of inferring the *phylogenetic tree*. In this method, taxonomic units are represented by their *DNA* sequences (most commonly, from the *mitochondrial genome*). The method looks for the tree that requires the minimum number of changes between the extant and inferred ancestral sequences. The outcome may be equivocal and, also, because the number of possible tree structures is extremely large, the optimal tree is frequently not found.

Perron–Frobenius theory Collection of results concerning eigenvalues and eigenvectors of positive (or non-negative) matrices and operators. Important assumptions include irreducibility (positive regularity) (i.e., a strict positivity of iterates of the matrix or operator). A generic result states the existence of a strictly positive simple eigenvalue dominating all other eigenvalues and of a corresponding strictly positive eigenvector. The importance of these results is that they lead to characterizations of the *asymptotic behavior* of iterates of positive matrices or operators, in terms of dominant eigenvalues and eigenvectors. Mathematically, $m_0M^i \sim \lambda^i \nu$, as $i \to \infty$, where M^i is the ith iterate of the positive matrix M, m_0 is the initial vector of states, λ is the dominant positive eigenvalue, and ν is the corresponding eigenvector. Results of this type are important in mathematical population dynamics, including the theory of *branching processes*.

pgf *Probability generating function*.

Phylogenetic tree The set of ancestry relationships between extant (contemporary) taxonomic or demographic units (species, populations, haplotypes, and others), usually in the form of a binary tree graph (at most three branches out of each node). The nodes of the phylogenetic tree represent extant and ancestral units, whereas the branches represent the intervals of evolutionary time separating them. Depending on the method of reconstruction, the graph may be rooted [i.e. having a uniquely defined common ancestor (and consequently, the direction of time specified in all branches), or unrooted (it is then sometimes called a network)]. The most commonly used methods of reconstruction are *parsimony*, distance matrix, and *maximum likelihood*.

Poisson process One of the most important *stochastic processes*. Random collection of time points (epochs) having the properties of complete randomness (the counts of events in any two disjoint time intervals are independent) and stationarity [the probability of an event occurring in a short time interval $(t, t + dt)$ is equal to $\lambda dt + o(dt)$, where, $o(dt)$ is small with respect to dt, i.e., $o(dt)/dt \to 0$ as $dt \to 0$]. The constant λ is called the intensity of the process. The number N of epochs of the Poisson process in an interval of length t has Poisson distribution with parameter λt [i.e., $\mathrm{P}[N = n] = \exp(-\lambda t)(\lambda t)^n/n!$, for $n = 0, 1, 2, \ldots$], and the time intervals T between any two epochs have exponential distribution with parameter λ (i.e., the density of distribution of T is equal to $f_T(t) = \lambda \exp(-\lambda t)$, for $t \geq 0$).

Population genetic models Models of inheritance, *mutation*, and selection of genetic material in populations of individuals. Classically, these models assume a constant number of individuals related to each other through common ancestry (Fisher–Wright model). Although very different from the *branching processes* some of these models can be approximated by branching processes (e.g., when an expanding subpopulation of mutants arises within the large population). Such a situation arises when some genetic diseases are studied.

Positivity In general, the property of being positive. A matrix is positive if all elements of the matrix are positive; it is positive regular if all elements are non-negative and some power of the matrix is positive. If the matrix is a transition probability matrix of a *Markov process*, positive regularity means that there exist paths between all pairs of states of the process. Similarly, if the matrix is the mean progeny matrix of a multitype *branching process*, then positive regularity means that any particle has, among its descendants, particles of all types.

Probability generating function (*pgf*) The function $f_X(s)$ of a symbolic argument s, which is an equivalent of the distribution of a non-negative-integer-valued *random variable* X. If numbers $p_0, p_1, p_2, \ldots$ constitute the distribution of random variable X (i.e., $\mathrm{P}[X = k] = p_k$), then the pgf of random variable X is defined as $f_X(s) = \mathrm{E}(s^X) = \sum_{i=0}^{\infty} p_i s^i$, for $s \in [0, 1]$. Use of the pgf simplifies mathematical derivations involving non-negative integer-valued random variables.

Quasistationarity State i_a of a *Markov chain* $X(t)$ is called absorbing if the process cannot exit i_a once i_a has been visited (i.e., $\mathrm{P}[X(t+s) \neq i_a | X(t) = i_a] = 0$). Under certain additional conditions, the probability of eventual absorption in state i_a is equal to 1 (i.e., $\mathrm{P}[\lim_{t\to\infty} X(t) = i_a] = 1$). Then, the only *stationary* distribution is the one that assigns probability 1 to state i_a. Because such a distribution is not informative, it is usual to consider a distribution, which is stationary conditional on non-absorption. Such a distribution, if it exists, is called the quasistationary distribution. Mathematically, $\tilde{\pi} = (\tilde{\pi}_0, \tilde{\pi}_1, \tilde{\pi}_2, \ldots)$ is the quasistationary distribution, if $\mathrm{P}[X(t+s) = j | X(t+s) \neq i_a] = \tilde{\pi}_j$ (all j) provided $\mathrm{P}[X(t) = j | X(t) \neq i_a] = \tilde{\pi}_j$ (all j). An example of a quasistationary distribution is the limit distribution of the subcritical *branching process* conditional on nonextinction.

Random variable (*rv*) Intuitively, a numerical result of observation which displays random variation. Mathematically, a random variable $X(\omega)$ is a function mapping the elements ω of a probability space Ω (space of outcomes of a random experiment) into the set of real numbers. For technical reasons, this function has to be measurable (i.e., the counter image of an interval through X has to be a measurable set of elements of Ω).

Random walk A time-discrete *Markov chain* $X(t)$, such that $X(t+1) = X(t) + U(t)$, where the integer *random variables* $U(t)$ are independent and identically distributed.

Recurrent state See *transient state*.

Renewal theory A branch of probability concerned with renewal processes. The renewal process is a collection of random time points (called renewals) such that the intervals between these points are independent identically distributed *random variables*. A special case in which the intervals between renewals are exponentially distributed is the *Poisson process*.

rv *Random variable*

Self-recurrence Consider a random (*stochastic*) *process* $X(t)$ evolving from an initial value $X(0) = x_0$ on time interval $[0, \infty)$. Suppose that at some time

t_0, the process is stopped and then restarted. Then, suppose that given the value $X(t_0) = x_0$, the continuation process on the interval $[t_0, \infty)$, which is a subprocess of the original process, is identical (it has the same distributions), as the original process shifted by t_0. A process with such property is called self-recurrent. Self-recurrence may be considered a rephrasing of a causality principle. It leads to recurrent relationships for a wide class of processes, including *Markov processes, renewal processes*, and *branching processes*.

Stathmokinesis An experimental technique in which *cell* divisions are blocked, ideally without damage to cells. Cells traversing successive phases of their lives are accumulating in the predivision state (mitosis). The time pattern of accumulation depends on the demography of the cell population and kinetic parameters of the cell cycle. Therefore, it is possible to estimate some of these parameters based on observed accumulation patterns.

Stationarity The *Markov chain* $X(t)$ is said to be stationary if its distribution over the state space is invariant in time (this distribution is called the stationary distribution). Mathematically, $\pi = (\pi_0, \pi_1, \pi_2, \ldots)$ is the stationary distribution if $\mathrm{P}[X(t+s) = j] = \pi_j$ (all j) provided $\mathrm{P}[X(t) = j] = \pi_j$ (all j).

Stochastic process Intuitively, a function of time with a random component. Mathematically, a family of *random variables* parameterized by time. It has to satisfy so-called measurability conditions, which prevent certain mathematical problems from occurring.

Transient state States of a *Markov chain* can be classified into transient and *recurrent*. For a recurrent state, the probability of eventually returning to this state is equal to 1, whereas for a transient state, there is a nonzero probability of never returning.

Type space A collection of possible particle types existing in a *branching process*. If there is more than one but finitely many types, the process is called multitype. If the type space is *denumerable* or continuous, the behavior of the branching process can differ considerably from the multitype case. An example is a *branching random walk*, in which the asymptotic behavior can be, for example, exponential multiplied by a fractional power function, which does not occur in the finite case.

wp *With probability* (common abbreviation)

Yaglom's theorem Result stating that for subcritical *branching processes*, there exists a *quasistationary* distribution, conditional on non*extinction*.

Yule process *Markov age-dependent branching process* in which a particle can have at most two progeny (the binary-fission process). An important class of processes because the *pgf* of the distribution of particle count can be explicitly found. Also, the Yule process frequently serves as a model for populations of proliferating *cells*, although by its definition it is limited to exponentially distributed cell lifetimes.

Index